WRITTEN IN STONE

Minerals collected and described by Roger Caillois

Curated by
Stefano Salis

Texts by
Roger Caillois
François Farges
Carlo Ossola
Marguerite Yourcenar
Stefano Salis

Franco Maria Ricci

Editorial Project
Edoardo Pepino

Art Direction
Laura Casalis Ricci

Curated by
Stefano Salis

Editing
Orsola Bontempi
Pietro Mercogliano

Graphic design
Laura Fava

Translations
Laura Bennet for Nuovo Traduttore Letterario, Florence
Laurel Saint-Pierre

The translation of the texts by Roger Caillois and Marguerite Yourcenar is by Barbara Bray and was originally published as *The Writing of Stones* by Virginia University Press in 1985.

Printed by
Grafiche Milani, Milano

Photographs
All the photographs of Caillois Collection are by François Farges

Other credits
Cornell University Library, Ithaca NY p. 34
ETH-Bibliothek, Zürich p. 37
© Sophie Bassouls. 2023 / Bridgeman Images p. 136

Cover

Agate probably Rio Grande do Sul (Brazil), thick slab, sawn and polished, 14 x 12 cm. Paris, MNHN, Caillois Donation, 1984-1985, INV. 188.98, inventoried as Désintégration, *probably an apocryphal name.*

© 2023 Franco Maria Ricci Editore
Labirinto della Masone
43012 Fontanellato (Parma)
tel. 0521 827081
www.francomariaricci.com
ISBN: 9791280294203

Acknowledgements

Catherine Rizéa-Caillois

Odile Felgine

Nicolas Bos and Marie Vallanet-Delhom (École Van Cleef & Arpels des Arts Joailliers)

The direction and the staff of Muséum national de l'histoire naturelle di Parigi

Sam Stourdzé, director of Villa Medici, Académie de France in Rome

Stefano Sampietro (Gallimard)

Leah Hildreth Stearns (Virginia University Press)

We would also like to thank Gallimard for granting us the rights to the texts by Roger Caillois and the Académie de France for granting us the rights to the text by Marguerite Yourcenar.

Contents

7

Mineral Mysticism
by Stefano Salis

33

Mineral Ecstasy.
Roger Caillois Writer and Reader of Stone
by Carlo Ossola

43

Roger Caillois's *Estro Poetico:*
of Minerals, Words and Dreams
by François Farges

65

The Writing of Stones
by Roger Caillois

129

The Man who Loved Stones
by Marguerite Yourcenar

132

Index of plates

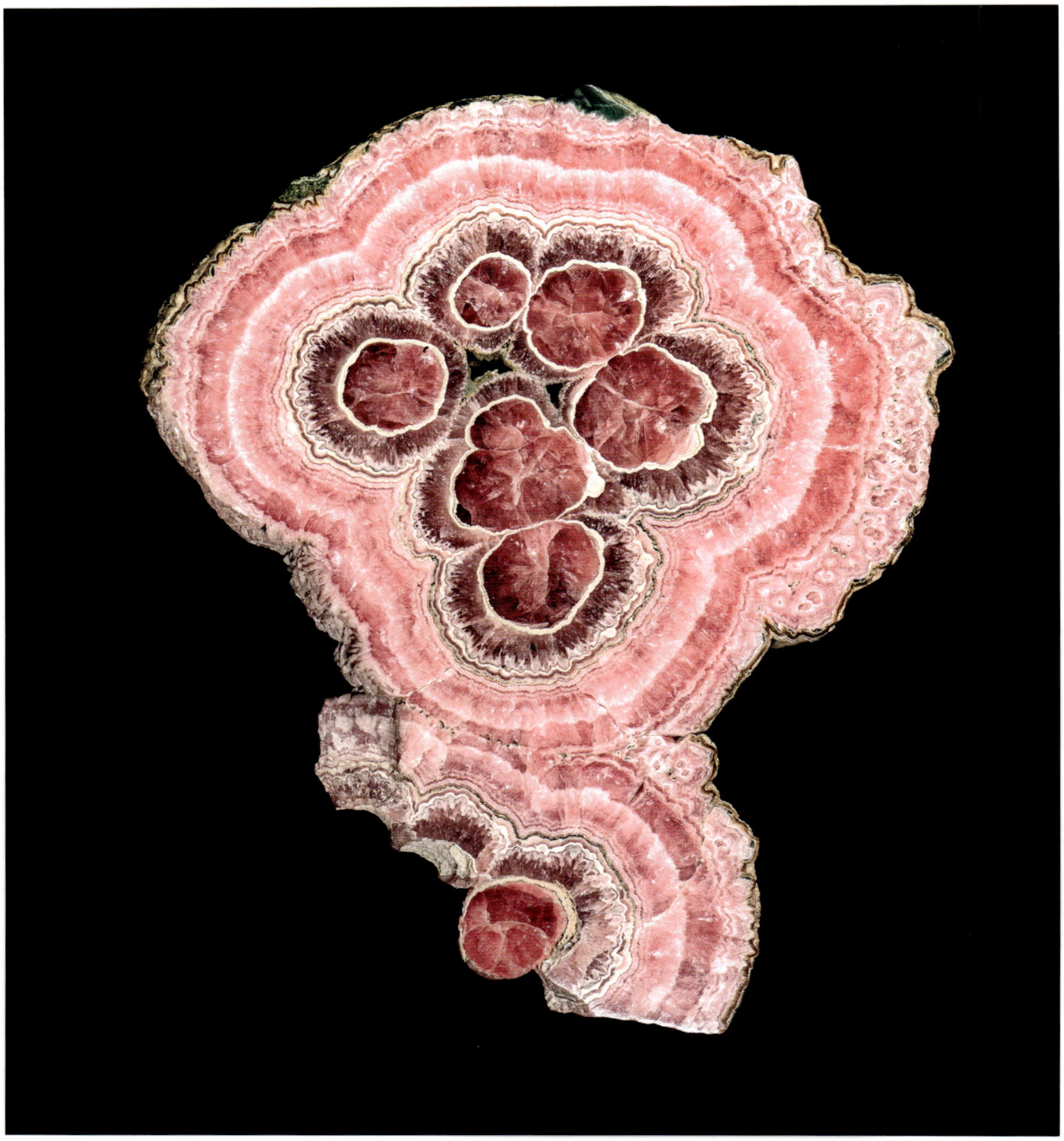

Mineral Mysticism
by Stefano Salis

An Immortal Signature

To find a way into the subject matter of Roger Caillois (1913–1978), and to better understand his diagonals, scientific and otherwise, within the mysteries of the cosmos and the ample space opened up by fruitful analogies, we must always think in terms of following a path in reverse.
It is necessary, that is, to work backwards through his life experience, which is the experience of an intellectual (though he strongly denied wanting to be one), in order to fully capture the ultimate meaning of his actions, writings, and feelings.
In Caillois's case, approximations are not enough, however fascinating. On the contrary, every movement must be meticulously interpreted, perhaps proceeding in a spiral motion or as if in a labyrinth. These two forms "of knowing" were very dear to him as a person who was highly erratic (and yet linear at the same time, which becomes clear when you consider the path he took and view it as a single whole). One must almost get to know his every word to be able to grasp how unequivocal his transformations are, how unified his digressions, and to see the profound coherence and "hidden recurrences" in what experience seems instead to scatter or link to ephemeral moments. Clearly, it is necessary to acquire the fixity of stones, whose existence far exceeds the hourglass of life, the works and days that the dust of mortal existence allows.
The flux of time, our position in the cosmos, the experience of a human life inscribed in a broader framework: these are all things Caillois suggests can be grasped through metaphors, recurrences, actions, transformations. Analogies. That is not to say that things must travel in a pre-determined, canonical, ordinary direction. If there is one thing we can learn from this highly erudite and singular writer (I use the term with great care: one could in fact describe him in many ways, all of them valid, not least of which is that he was a "man who loved stones", to quote Marguerite Yourcenar, who succeeded him as a member of the Académie Française), it is that the appearance of a progressive path is just that – an appearance.
In the case of Roger Caillois, it is no accident – far from it! – that he chose as the metaphor for his own existential path the river Alpheus. It is the title of his spiritual biography, which is certainly not a mere list of events but a distillation of the sense of the tortuous path he pursued. He chose a river – making use of a myth that has been interpreted in many ways – that sprang forth in Greece and was named after the river god Alpheus. Alpheus, the story goes, out of love for a nymph who had fled from him, and thwarted by the volatility of the ruling gods, threw himself into the the sea to reach her, without allowing his own waters to mix with those of the sea. Then he retraced the path that would lead back upstream to her: to the Aretusa spring in Sicily. Having taken a path, he returns to his origins: from estuary to source.
It is therefore necessary to take note of certain events that occurred in the writer's life in the crucial year of 1978, that of his death.
He was a man who received honours, led an intense life, and who worked as both a writer and researcher of apparently marginal matters, continually travelling the world as an official for UNESCO: these clues may help establish some fixed coordinates. By tracing a spiral motion, we may go back and try to understand, *ab origine*, his movements, his reasoning, his words, and his stones.
Words and stones. This is the dual universe in which we must move.
A first clue. In November 1978, almost exactly a month before his death due to a sudden and incurable cerebral haemorrhage, Caillois received two journalists from the *Nouvel Observateur*, Héctor Bianciotti and

Jean-Paul Enthoven, at his home (in which he had a magnificent collection of stones, not so much a display for others as an eternal *memento* for Caillois himself).
A dense and highly meaningful interview was published a few days after the writer's death, becoming both a kind of final testament and a paradigm; or rather, it acquired a correlative value. This was not just owing to the accidental circumstances of his death, because Caillois, who chose every single word with great care, like a philatelist selects stamps (Caillois himself used the image in the interview), had unfailingly expressed himself with great precision. What's more, 1978 had been a very eventful year for him: in a short space of time, he had published books that marked, like an ominous sign, as Jean Blot has noted, or perhaps foreshadowed, his imminent demise. *Le fleuve Alphée* (The River Alpheus) had come out in March, together with two other key texts. One was *Trois leçons des Ténèbres* (Three Lessons from Darkness), a title that resembled but did not parody a composition by Couperin, differing from it imperceptibly, and which introduced a word, "darkness," that is of great importance for Caillois. The other was *Le champ des signes* (The Field of Signs).
Caillois had been a member of the Académie Française since 1971 (he memorably described his election, replacing the historian of ancient Rome, Jérôme Carcopino, as "a double misunderstanding," and gave a memorable *lectio* in the presence of the Immortals that took place in candlelight due a sudden black-out – a Luciferian invitation, someone murmured, which Caillois must have engineered and indeed may have slyly embraced) and received one award after another in that period, which was the time of his life. In 1978 alone, *Le fleuve Alphée* won the Prix Marcel Proust soon after it was published and in September he was also awarded the Grand Prix national des Lettres for "his body of work." Two more idiosyncratic works appeared that same year: *Rencontres* (Encounters), an overview of a literary and critical life (as he described it to Bianciotti) and *Approches de la poésie* (Approaches to Poetry).
At the very end of the interview, in the final question, is the first key point we should focus on.
"What image would you like people to have of your work, of you?" the interviewers asked. Caillois knew he was about to say goodbye to his guests, and so once again he chose his words with great care. He replied: "Perhaps and *exclusively* [that] of a poet." The emphasis is mine. The function of that adverb (which contradicts one of the rules Caillois lived by: brevity and precision) was to lay down a condition for the reading of his work. But he went on to specify: "Of a poet that dares to say: I speak only for myself, but it is as if everyone, in my words, were expressing themselves like me and daring to say: I address an invisible interlocutor, but in such a way that everyone can share the illusion that my words are addressed to them alone."
And then came the decisive thrust: "Impersonal confessions of a hidden shadow to anonymous shadows…"
These were the last words of the interview; published posthumously, they were – literarily – the last living words of Roger Caillois.
Biographies are always very ambiguous: they interpret, *a posteriori*, and try to give substance to the experience of others, reconstructing (at times successfully, at others much less so) facts and events in a way which even the subjects involved may not have managed to plan or execute with the same inner coherence. Unconscious acts, in the light of overall experience, acquire different refractivities and colours: a rainbow of meanings serve to paint with clarity and sharply delineated lines the tangle of lived experience, both mental and material. So, Odile Felgine, the author of the only – and admirable – biography of Caillois, relates an episode which demonstrates – I was about to say incredibly, but with Caillois that word

loses its power – the cosmic energy of a spiral.
I will leave it to Giuseppe Zuccarino, who, together with Roberta Coglitore, is the leading and subtlest Italian interpreter of Caillois's work and legacy, to describe how it ends.
In the final months of 1978, Caillois was very weak and ill, and he was not even able to attend the award ceremony for the European non-fiction prize that he had won. However, he did not want to miss out on visiting an important mineralogical exhibition held each year in Paris. As he struggled along the corridors, his eye fell on an agate that seemed to him to depict a current crossing the sea and returning to its source. He recognized it.
It was the river Alpheus, as he had recently described it in his autobiographical work. He was just in time to buy the stone; not long afterwards he had to be rushed to hospital after suffering the cerebral haemorrhage that would lead to his death shortly afterwards.
The anecdote – which is almost too good to be true – was the perfect seal on the life of an ambitious, restless, and extremely erudite writer who, after having learnt (late, but admirably) the difficult art of "reading" and transcribing minerals, had the good fortune to see inscribed in one of them the very image that he felt encapsulated his destiny better than any other.
This is all very touching, without a doubt, and it matters little if it is slightly romanticized.
Yet there is more. On December 26, 1978, after a religious ceremony in the church of Saint-Séverin in Paris that was very brief (Caillois was not a believer and had not the slightest inclination in that direction), the writer was buried in Montparnasse Cemetery. Another, equally paradigmatic, story begins here, which Caillois had certainly foreseen. Indeed, his talents as a director were hardly hidden, and we have (indirect) testimony from the author himself, who, back in 1965, had written a singular narrative entitled *Récit du délogé* (Story of the Dislodged). The analogy has been subtly investigated by Roberta Coglitore. *Some stories are told in the first person, but under a false identity,* the scholar explains. *A first person narrator and character, who only in part or through displacements can be assimilated to the life and character of the author (for example, Récit du délogé attests to Caillois's passion for stones, a passion so strong it grips the imagination of the 'I' narrator to the point of making him believe he is 'inhabited' by a stone, while in the course of the text that same 'I' presents himself as a sales assistant in a fish tackle shop, totally removed from the reality of the author's life.*
We must be careful here: in that story, which came before all the books about stones and their cosmic significance, Caillois imagines that the main character realizes a small stone similar to "a cuttlefish bone" has "gestated" inside him, just above the pubis in the lower gut. And that this mineral excrescence, very similar to a seashell (the importance of which will become clear), grows and "displaces" the person, taking their place definitively. Fanciful, but ultimately not so very much.
Let's return to the funeral, and to the story it relates. Once again, I will leave the telling to someone else, to André Chastel, who is much better qualified than I am or indeed any of us are.
Not only was he a leading critic and art historian, and an expert on the work of Caillois, he was also a long-time friend, having met Caillois back in 1930 when the two went to school together.
In early January 1980, exactly a year after his friend's death, Chastel published a poignant article in *Le Monde*. More than just a tribute to a friend and an interpretation of his intellectual career, it formed a kind of amicable dialectic which, as a recollection of those glorious years, could not avoid certain details. And Chastel had to start from the very end: from the mineral legacy of his friend, his tombstone, as the title of the article suggests.
In Montparnasse Cemetery, in the

area reserved for literary figures, a tomb has just been erected which stands out from all the others. [...] Anyone who knew Roger Caillois cannot possibly fail to recognize the 'monument', which suits him perfectly. We are talking about a bare slab of pink and grey Alvernia sandstone, with no inscription at all: and within this prismatic stele, as in the reliquaries of times past, are the precious remnants of a gigantic ammonite fossil (the work is by Nicolas Carrega), Chastel writes. He then recalls an anecdote: "'How wonderful, not even his name is there,' he [Caillois] had said in front of the tomb of Grand-Bé." The reference here is to Chateaubriand, buried in an anonymous tomb facing the great Atlantic Ocean. Or rather, with an inscription added years later, which Caillois must have known about: *Un grand écrivain français a voulu reposer ici pour n'y entendre que le vent et la mer. Passant respecte sa dernière volonté.*

Chastel continues: *In one of the prose poems that he wrote in his final years, we read that he envied the name of Nobody used by Odysseus and that of Nemo, Jules Verne's captain.* Caillois had, after all, travelled widely in the world like the former (when he worked for UNESCO) and, as Chastel notes, had descended *into the mysterious depth of nature like the latter, examining the mimetic traits of insects and the singularities of minerals. He did everything in his final years to lucidly enter the anonymous realm, once his work was done. The only fitting tribute is to ask questions together with him.*

Precisely because he knew him so well, Chastel grasped the essence of Caillois. Later in that fine article he writes: *Initially it was the law of Valéry: not much awareness, and the law of André Breton: not much reality. How it was possible for these two demanding rules to coexist, fight and intensify each other is evident in this work, where the tone of* intimacy *is made possible by the* growing impersonality of the content. *It was necessary to work in the sphere of the marginal, the absurd, the scandalous, the strange, to go into disturbing points of intersection, to try out and to trace the unusual. Later on, we will better understand what an enormous legacy this attitude assumed, what a continual exercise between dream and positivity it imposed on this rigorous, playful spirit. In a universe where everything holds up, it should be possible to demonstrate that the lyricism of Saint-John Perse* [the encyclopaedic poet long admired by Caillois, who wrote many studies about him] *is equivalent to the splendours of the 'septaria'; as the poetic metaphor brings us up somehow in the layers of evolution. But it is not a question now of discussing these individual proposals. The tomb in the cemetery of Montparnasse reminds us of something simpler and more tragic.*

The words in italics, "intimacy" and "growing impersonality of the content" echo the final words uttered by Caillois in his interview with the two journalists, his testament as an ephemeral, anonymous human being for other human beings.

Further: "The strange observations of 1970" – here Chastel is referring to the preface of *Cases d'un échiquier* (Squares on a Chessboard) – "signal a slow deceleration marked by a fondness for minerals, their crystals, and their images in which he considers disquieting similarities and a series of intellectual affinities: *Pierres réfléchies* (Reflected Stones), 1975 and *Rencontres*, 1978; and undertakes a literary self-examination: *Le fleuve Alphée*, 1978".

Then, to conclude: *Sumptuous minerals become object, symbol, ultimate pretext, and proof of the indivisible unity of the world, where the embroidery of consciousness strives to find immemorial and silent unfurling, as if it no longer existed, in this hidden complicity, only in order 'to complete the order of the universe on a tiny point'. This point disappears almost immediately, and it is the silent slab, the anonymous stele, the stone, which has the final word. And, quite marvellously, thanks to what must be called art, without any betrayal.*

Poetry, being a poet, the river Alpheus, the silent tomb without a name: just a stone and an ammonite, the crystallization of a living being which was stone and returns to being stone. I return now to that final interview: "The imagination for me is nothing other than the extension of matter, and I believe that poetry is not a purely human phenomenon and that it is not a phenomenon deriving from language." We are at the heart of Caillois's final vision here. Having arrived at a supreme awareness of his work and philosophy (a term he detested: he considered philosophy, thought, books nothing less than a vice, a disease...), he could finally reveal what he meant deep down with his work. Immediately after these words, he moved on, fatally and inexorably, to the realm of stones.

If, for now, I concern myself only with describing stones, I do so to show that inside these stones, and in the manner in which they find their form, there is a kind of reduction or miniaturization of all the things that exist in the world. The imagination is under the illusion that it embellishes in its manner, and yet the interior of a stone proclaims that rigour and drift do nothing other than work an immortal, invariable canvas. The same mysterious rules govern inextricable mental freedom and mineral gravity.

On the final page of *L'Écriture des pierres* (The Writing of Stones), so as early as 1970, Caillois foresees the cosmic bond that would then be reflected, incessantly and endlessly, in his words and on his tomb. He describes the beginning of life, and how it is recorded by the fixity of stone:

Ammonites of every size, from the lentil to the millstone, impose their cosmic spirals everywhere [...] Compact and insensible matter has replaced the other kind in its last refuge, taking over its exact shapes, running in its finest channels, so that the first image is set down forever in the great album of the ages. The writer has disappeared, but each flourish – evidence of a different miracle – remains, an immortal signature.

Written in Stone, a "Magical Resin"

I speak of stones: algebra, vertigo and order; hymns and quincunxes; stings and corollas, the threshold of dreams, catalyst and image. Of a stone that is the cascade of hair, opaque and stiff, of a drowned person [...]; of stones that are smooth paper, non-combustible and sprinkled with uncertain sparks. [...] I speak of stones that existed before life and will remain after it on the cold planets. I speak of stones that do not even have to wait for death and that have nothing other to do than let sand, rain, backwash or storm, and time, slide over them.

This quote was necessary, first to give readers a chance to experience the quiver of excitement aroused by Caillois's enchanting prose (to which we will return), but also to allow the author himself to introduce us to the solid, yet mystical realm of his collection of stones.

Caillois, an eclectic intellectual like no other in the last century, endlessly fascinated by the "natural fantastic," a fearless dreamer who knew the vertigo of literature, the rigorous geometry of science, and the disjointed, omniscent, and all-encompassing world of myth, wrote these lines in 1966, when he decided to systematize his thinking about these fragments of infinity that had been turned into matter and gift it to us, for our great benefit, as inhabitants of a large, remote spherical stone in the middle of the cosmos.

The origins of the book date back to 1962. We know from his letters that Caillois had begun to "systematize" his "dark," disquieting, cosmic, and profoundly human thinking. It would take him several years, but from then on, the mineral universe would be the preeminent – I don't want to say exclusive – focus of his intellectual, that is his poetic interest.

Over the years he collected stones of every type, size and provenance. In 1965 he cast aside the timidity of the collector and showed a selection of his pieces in the Parisian gallery of Katia Granoff. In the same year he published one of his emblematic works, perhaps

his last great book "before the stones", *Au coeur du fantastique* (At the Heart of the Fantastic).
Stones kept him company at home and their silent narrative kept him close. They attracted his attention and suggested to him a higher experience of life – yes, of life, cold, "dead" stones though they were. Awaiting frozen time, they accompanied him, unaware of our existance on Earth – agate and chalcedony, paesina stones and jasper, ruin stones and "dream stones," as the Chinese call those with unintentional figures traced on their smooth surfaces. Works of art of a hidden god, manifested in an imperturbable and – perhaps – not accidental beauty.
There are stones that do not end up in a natural history museum, as seems obvious, but instead compete with human painting, sometimes surpassing it but always, in some way, foreshadowing it. What Caillois "sees" in stone is the very essence of literature and, much more than that, of our destiny and way of being human: beings who dream, and who are capable of imagining, of deriving, from cold matter born of violent and ancient compressions and eruptions, opportunities for wonder and amazement, contemplation, and wisdom.
Caillois "titles" stones, conferring on them the power of narrative; he imagines metaphors and stories emanating from those fantastic inert slivers and records everything with an encyclopaedic culture nurtured by consciousness-forming myths and instincts; as when, with a poignant image, he notes that the classics already saw black, vitrified, translucent obsidian as a mirror not for images but for shadows. As if by magic, mineral writing and alphabets appear here and there in his stones – works of pure draughtsmanship or exercises in sublime "resemblance" or natural prophecy – and a whole lapidary encyclopaedia is conjured in the tumult of casual lines: "a systematic repertoire of signs, which are not necessarily letters but do, however, constitute coherent and associated symbols," as Caillois wrote in *Malversations* (Misappropriation).
Caillois arrived at stones, and this is the great thing about the whole matter. And if there is one point that he does not fail to emphasize in his autobiography it is that his trajectory was inscribed in a path following basic lines. He had a "rural" childhood of a kind not uncommon for a child born in the French provinces at the outbreak of the First World War (a feature Yourcenar was quick to grasp in her decisive way). He learned to read quite late, but that would be his downfall: he became a voracious and inveterate reader, a model secondary school pupil, an early Surrealist, founder of the College of Sociology, a writer once and for all.
He was an essayist and mythographer who became totally immersed in and dominated by books. But in his later years, thanks to "antidote books" (as he himself explained), he came to the end of this long bookish "parenthesis" and discovered the paradise of minerals, at the same time recovering the best of his childhood (which Chastel did not fail to underline).
In *Le fleuve Alphée*, Caillois writes:
Just as the river Alpheus, I suppose, was saved by the memory of the first phase of its course, I will have been by that of my childhood. Just as it preserved the waters that wanted only to mix with those of the sea, I instinctively clung to the objects that chance had placed in my path, like those rare pictures and poems which a mystery I could not solve made me link to objects rather than painting or poetry. Consequently, poetry as a whole, the culmination of verbal images, of inextricable emblems, the prevarications of mimetic animals, the dangerous, splendid and indistinct condition of plants, has always offered me silent complicity. I have only ever half-spoken of it, and only to warn of its magic.
Here, now, is the key passage: *Endlessly enlarging the circle of a solidarity that took me as far as possible away from myself, I finally found the longed-for reward in stones. Little by little, they showed themselves to be a gigantic*

album. Situated at the extreme of silence, they lay at the same time at the opposite pole to man and thought. I guessed that they contained some possible transformations of matter in their impassive, enduring mass, without excluding anything, not even sensibility, intelligence, imagination. At the same time, totally mute, they seemed to me to challenge books and to convey a message outside of time. [...] Stones, supreme archives, which carry no text and give nothing to read...

This reflection, significantly included in the first part of *Le fleuve Alphée*, ends with an ellipsis. Caillois is too accomplished a writer to grant himself the luxury of being imprecise. No. That ellipsis alludes to the ironic (or rather, self-ironic) meaning of the sentence and of Caillois's position (the final decade of his life was spent composing a supreme mineral poem); it holds the promise that he himself would contradict that statement, and brilliantly so.

Roberta Coglitore observes: *Beyond his own personal experiences or themes of inquiry, Caillois situates the most important boundaries of his autobiographical journey around his tormented love for literature and writing. As we read in* Le fleuve Alphée, *a long parenthesis of his life – from which he consciously and deliberately excludes the literature he worked on before he became involved with Surrealism – concluded at the end of the 60s with the discovery of the centrality of the mineral world, which enabled his reconciliation with a new literary writing, this time 'en pure perte.' In his autobiography, he writes: "[les pierres] m'ont rendu au contraire une raison d'écrire, au moment où je doutais de celles qui, jusque-là, m'en avaient persuadé." Human hopes of eternity, nurtured by faith in the lasting value of writing, which had motivated Caillois in his research, became, on the contrary, once he had moved past the parenthesis, an 'acte d'allégeance' with the non-human dimension, to the point of moving towards a complete assimilation with the nature of stones. Stones, 'de l'origine et de la fin,' which date back to geological eras that precede the appearance of humankind and will endure after its inevitable dissolution, also mark the initial and final phase of Caillois's own life.*

In his interview with the two journalists, Caillois used the word "describe" when talking about his activities involving stones. What does this word mean exactly for Caillois?

To answer this, we can turn first to a short critical essay, again by Roberta Coglitore, before looking directly at the strategies employed by Caillois as a means to an end.

Coglitore writes: *Caillois dedicated the last twenty years of his life to his passion for minerals, in an attempt to establish a generalized poetics in which he could situate, on the same line and without interruption, the 'physical, intellectual and imaginary world', as he himself wrote in* Le champ des signes. *In the general project of the diagonal sciences, which cuts across knowledge categories in search of analogies between manifestations of nature that do not find answers in traditional disciplinary fields, Caillois goes so far as to hypothesize an anthropological decentring. The profound correspondences between phenomena scattered across the universe are the demonstration of a harmony but also of a beauty, if not a full-blown aesthetic project of nature, which humans can do nothing other than accept and acknowledge as the millenary and perfect riverbed of forms, within which they delude themselves that they are manifesting their art in an original way. From Caillois's perspective, in fact, the works of painters are 'a human variant of butterfly wings' and not vice versa. In Caillois's consummate descriptions, figured stones acquire the status of artworks and become the locus where nature's designs are completed. Without the reading of the designs – without, that is, recognition of the bands of colours, lines, planes, and depths shown in their sections and the relative composition of visual or mental images, a recognition that can only come from interaction with the material by sectioning, shattering, or simply uncovering it – the figures on the surface*

of minerals would remain concealed for ever in the depths of the earth, and would therefore have no meaning.
Obviously, all this can be accepted. But how does Caillois carry out this strategy? What do these critical comments mean? Where does one start? What do these "consummate descriptions" consist of? Well, here is an example from Caillois himself, from *Pierres* (Stones, 1966), which Giuseppe Zuccarino also regards as one of the best examples of the author's way of proceeding.
As you read the passage – and I am sure everyone will admire and be moved by it – note how Caillois slips from the scientific or "neutral" into the poetic and the lyrical, with a diagonal that could only be produced by a talent and a spirit like his (whose debt to Surrealism is evident, though he left it behind completely). The pleasure of reading the passage is multiplied for those who, like me, have transcribed it: rewriting the words one by one (albeit in translation) perfectly renders the essence of Caillois's operation. It makes clear the method and the substance, the poetic artifice and its essence, the prison of words and their marvellous opening onto the cosmos:
The nodules of agate are coarse grey and frankly unattractive balls. They have to be broken to discover the spectacle they sometimes conceal within; mostly though they are just dark, opaque material barely different from that of any type of silica."
Then the poetry begins:
"But sometimes [there are] whimsical markings, parallel veins whose meanders, before their course ends, have followed, not without accidents and second thoughts, the outline of the crust enclosing them; more rarely, mineral festoons, points of lace, sketches of pistils and sprays of pollen; explosions of chrysanthemums, frozen pyrotechnics in a petrified night; the transparency, long promised and long deferred, which suddenly rises like a spectre or meteorite; lachrymatories with their tears of a magical resin, the white star impressed as if on the forehead of a horse; glittering windows, the colour of cyclamen or wild rose; drapes and curtains – hanging in the midst of the stone on invisible hooks –, the folds of which fall solemnly, far away from each other like impassable mountains, draperies of Northern Lights; banal ditches, as deep as buckets of bitumen, in which are deposited the dregs and waste of ancient quintessentialized fluids; ribbons displayed in the slender thickness of their hem, with a marvellous train stretching out wavily in an opacity of honey or burning amber or pale blue lye, like the wings or fins of sea bats, or like lines.
The rush, the elevation, the lyrical dizziness is complete – a single, long sentence, a tumultuous and moving crescendo where the imagination burns and plays alongside the language. Poetry.
Caillois knows that he must now decelerate, to bring us back to human levels and to let us breathe and make peace with this overpowering demonstration of literary prowess which still perfect adheres to the represented object.
And he ends, once again, in a though-provoking and inspiring way: I*n no way am I considering everything that a lost or joyful imagination, which projects a people of simulacra in it, can identify in these shadows: towers, gods, monsters, caravans advancing beneath the palm trees of oases. I look at these designs as they were at the dawn of the ages, when nothing existed except for them, when matter, on the point of hardening, invented the first equilibria and, adverse to any discipline, avoided order and symmetry where necessary – to experiment with them elsewhere.*
A propensity for marvel, challenging the more or less concealed warning that Caillois himself hints at when speaking of stones: that by subjecting them to extensive explanation one somehow undermines their secret mystery, their silent poetry. This is the only way that we will learn, with Caillois, to read – sheet after sheet, stone after stone – their message, our transient humanity.
"I speak only of bare stones – fascination and glory – that both conceal and confide a slower, more immense and more profound mystery than the fate of an ephemeral species."

Pure Loss. A Mysticism without Ecstasy

In Marguerite Yourcenar's marvellous panegyric about Caillois, there is a point (whether rhetorical or not) on which she dissents from the writer she succeeded as a member of the Académie Française (she was the first woman to be elected). "He did not write in pure loss."
The expression, magnificent and evocative in its conclusive absoluteness, is, once again, the final sentence of one of the most crucial moments of *Le fleuve Alphée*. Caillois is talking about the "bubble" in which everyone is immersed, and touches on his own personal case. He says he is "a champion of anthropomorphism in reverse. I search in man for the outlet of hardened energies or of weaknesses and temptations that existed before him".
In this research, and in abandoning the workings of his own inner life, which he judged to be of very little interest, Caillois outlines the end of the human species and the attraction he feels for places where human presence was most remote and distant: physical places where the laws of aridity (the topic of one of his final pieces of writing) dominate and human presence is faint if it exists at all.
Immediately afterwards, alongside these considerations, however, the spirit of the writer ("as I am not ashamed of being," he casually drops into one of the final lines of *Pierres*) re-emerges forcefully: "I did not become reconciled to writing until the moment I began to write with the consciousness that I was doing it in any case in pure loss."
Words and stones: now they return. We really are a step away, in the "heart of the stone", and Caillois, not even a page later, in the final chapter of his autobiography-testament, decides to reveal the decisive link, the fulcrum of his strategy. His essence, as a man, and as a poet.
I try to give my sentences the same transparency, the same duration and, if possible – why not? – the same sparkle as stones. At the same time my early mistrust of literature and the hesitation I voiced in the moment in which, in vain, I reproached a certain rather too readily vertiginous poetry, this mistrust, these scruples were soothed by the fact that I am sure I am not lying and because it's a question of insensate minerals. In a word, I feel justified in the singular undertaking of looking for a new poetry in exactness.
It was Jean Starobinski, an attentive reader of Caillois's final work, the *Trois leçons des Ténèbres*, who recognized his saturnine nature, an almost pacified melancholy, a self-negation (in the Luciferian spirit), a "serene joyfulness" (this is Caillois himself speaking). And in one of his "exercises of admiration," Emile Cioran recognized the nature of the "failed" mysticism of the writer's construction.
"Mysticism" is a word to use with caution, as it recurs in Caillois and could give rise to misleading interpretations. There is no doubt, in fact, that the mineral mysticism evoked by the writer, always, of course, with regard to the sensations achieved in the lyrical (and cognitive) heights of his descriptions, can lead to a heightened awareness, a new way of transfiguring and understanding the unity of the cosmos, the true fixed point of Caillois's whole inner life.
When I look at stones carefully, sometimes I apply myself, not without naivety, to guess their secrets. I lose myself in imagining how so many enigmatic marvels could have formed, the product of laws that they rarely seem to violate, as if they were the result of a tumult or of a celebration that announces their way of being. I struggle to picture them in my mind at the burning instant of their genesis. I am then gripped by a very specific kind of excitement. I sense myself taking on parts of the nature of stones. At the same time, I bring them closer to my nature due to the unexpected qualities that I attribute to them during speculations that are at times precise, at others uncertain, in which the web of dreams and the chain of knowledge are combined. Fragile and perhaps necessary edifices ceaselessly rise and crumble there. The metaphor helps (or corrupts) the syllogism; vision feeds (or misleads) precision. Between the

fixity of stone and mental excitement, a kind of current flows, in which I find, for a single but memorable moment, wisdom and comfort. I can almost see the possible seed of an unknown and paradoxical kind of mysticism.

We have reached the central point here. Is it not a perfect description of what medieval mystics, Eastern sages, fiery shamanic star dancers, have, down the centuries, experienced when contemplating something larger than the finite nature of man? Between "the web of dreams and the chain of knowledge" – Caillois could not have identified more clearly the mystic essence of his experience.

But then he immediately clarifies: *Like other mysticisms, this too would lead the soul to be silent for half an hour, it would cause it to dissolve in some inhuman immensity. But this abyss would have nothing of the divine. It would be solely and purely matter, the active, turbulent matter of lava and fusion, of earthquakes, orgasms, and great tectonic events; and the motionless matter of the longest quietude.*

Quietude. Nullification in immobility, greyness even, an essential word with which the story of the fisherman ends, the fisherman "displaced" by the stone growing inside him. A fossil that is lost in the bare stone, nameless, in the silence of the cosmos.

I'll end with a final, definitive quote: *Stones, present at the origin of things, become confused with things themselves. And there is nothing human that is not immediately extraneous to them. They will survive in sideral space after the universal and its inevitable dissolution. The traces of a fleeting parasite will be nothing but traces in the texture of the stones. Fossils for no one. A perfidious demon suggests to me that describing them is equivalent to reflecting their eternity for an instant. I look at them for a long time before specifying their form, enumerating their features; I return to them many times, to the point of fatigue, to check a detail or to place them at another angle. In rare moments a powerful daydream suspends my vigilance. It prevents me from governing my thoughts. I allow images to assail me, to prolong the stone; I do not allow them to be unfaithful. I don't expect to arrive at any kind of revelation. It is simply a stone that is always there before me, that does not reveal the smallest secret to me. I don't even have the impression of lose myself in a distant absence. I feel a calm happiness. I feel rewarded, without on the other hand knowing for what effort or what virtue. I receive confirmation of a knowledge that I did not know belonged to me.*

A *knowledge* that arrives from the cosmos and returns to it: not scientific, not mystic knowledge. But perhaps the more important and deeper knowledge of a person who knew himself to be, above all, and indeed *exclusively*, a poet.

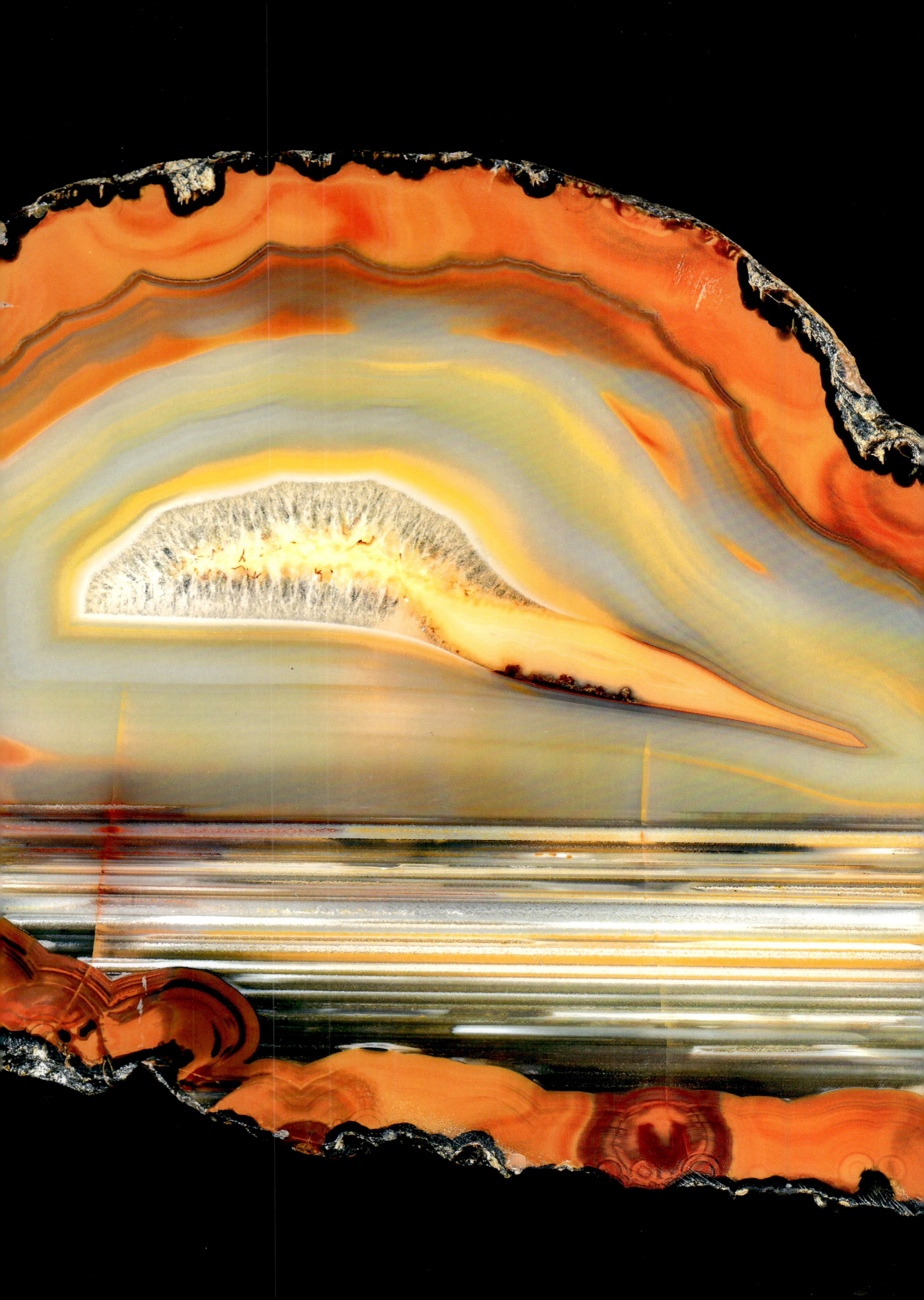

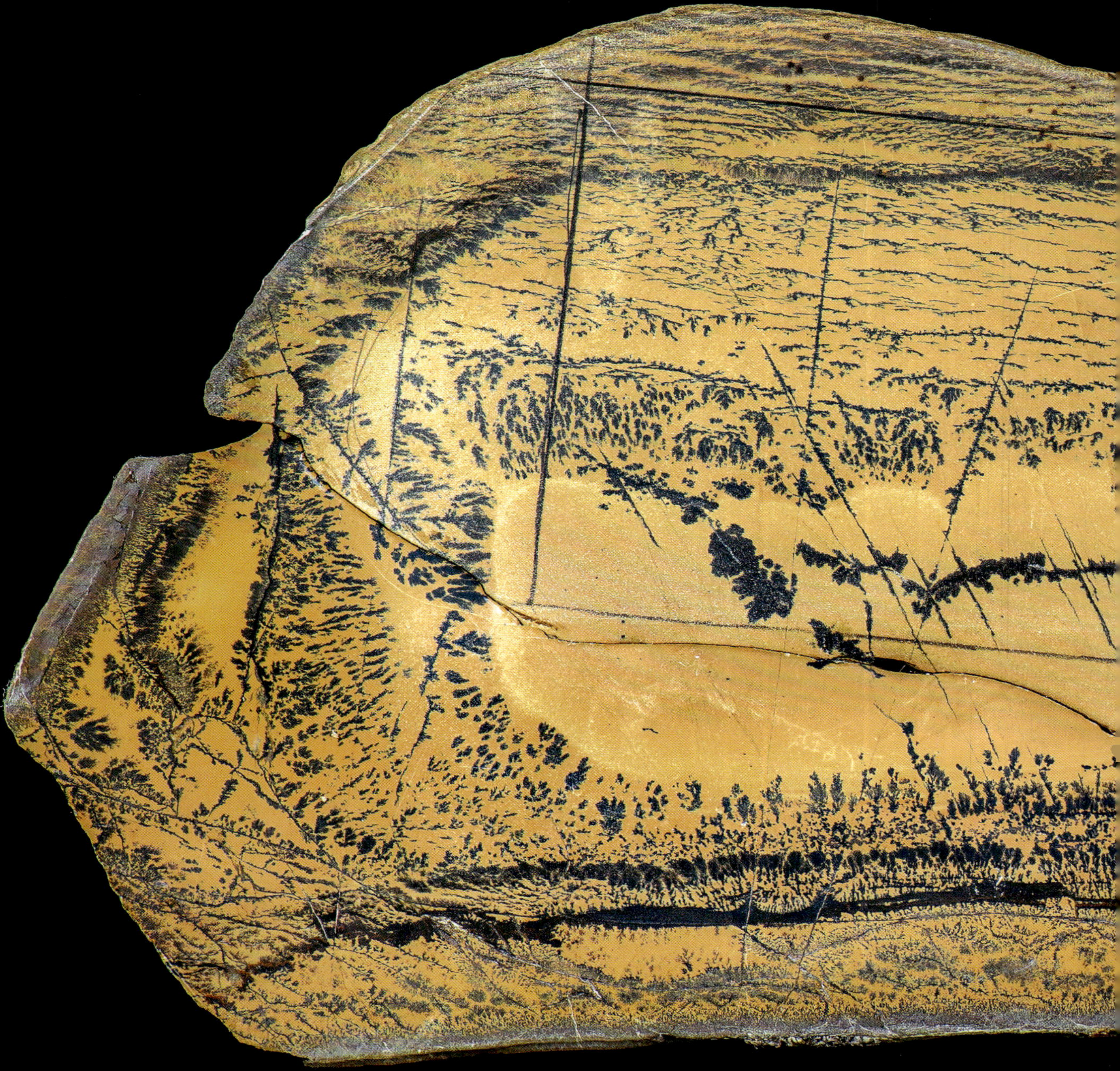

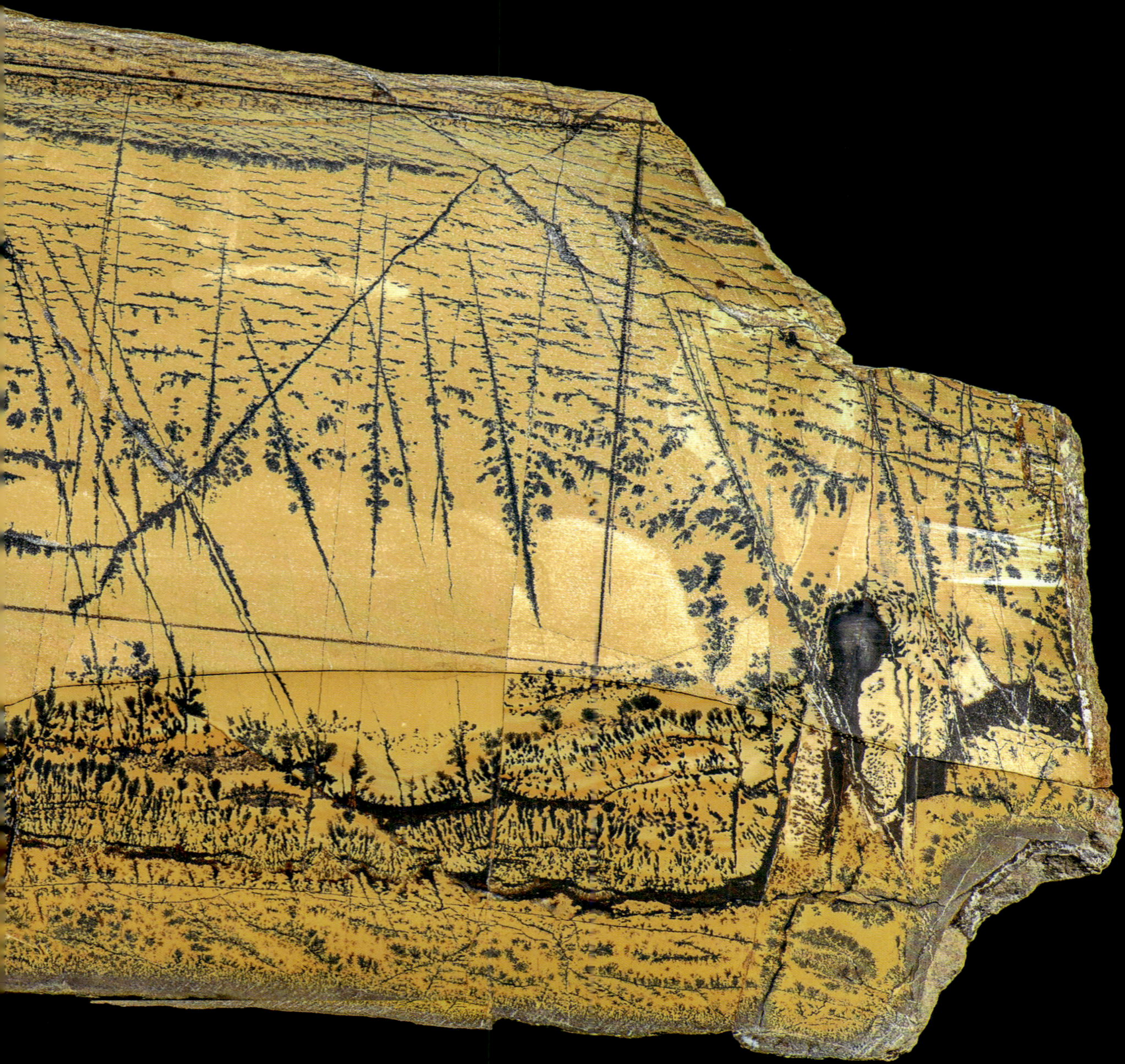

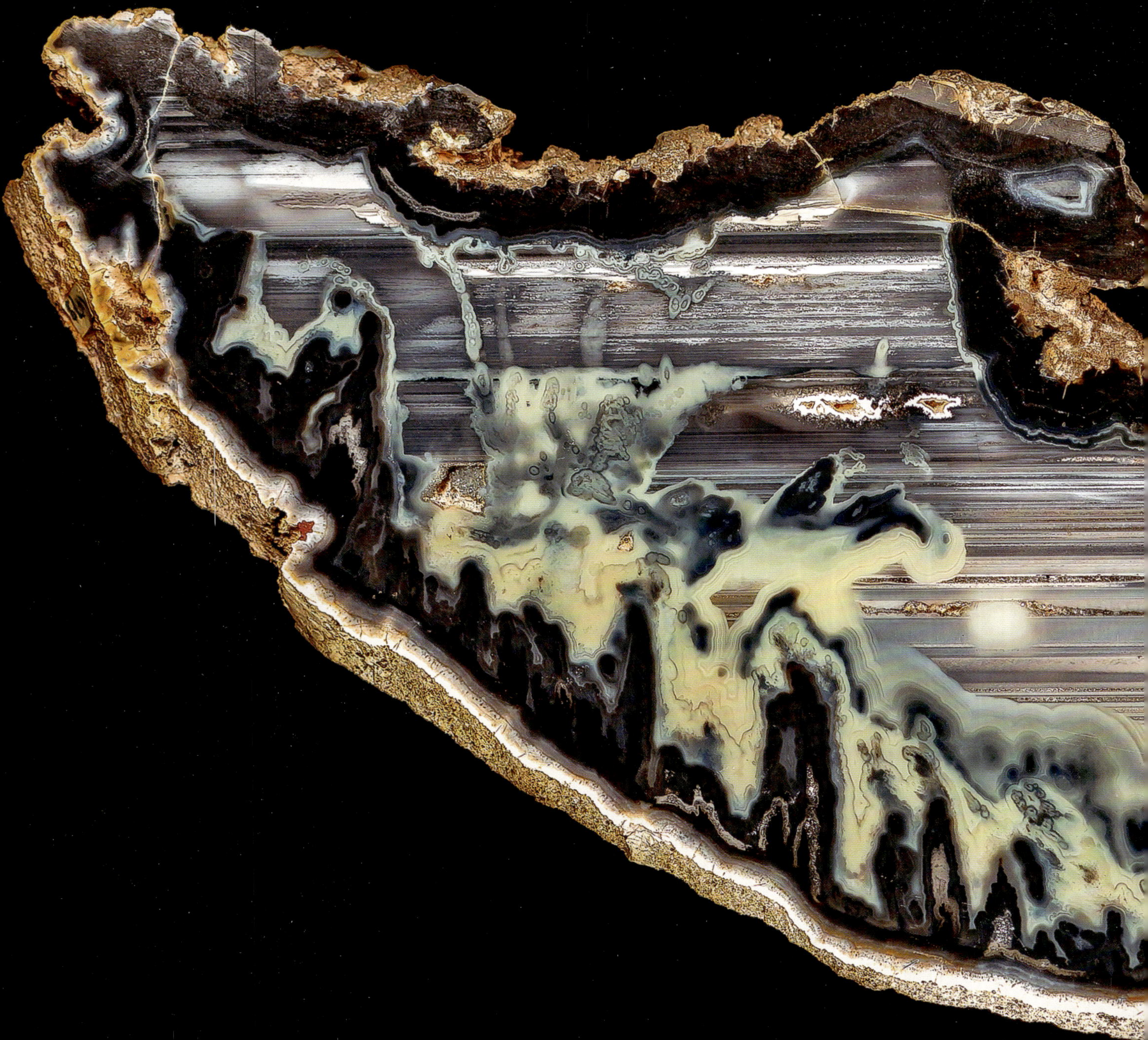

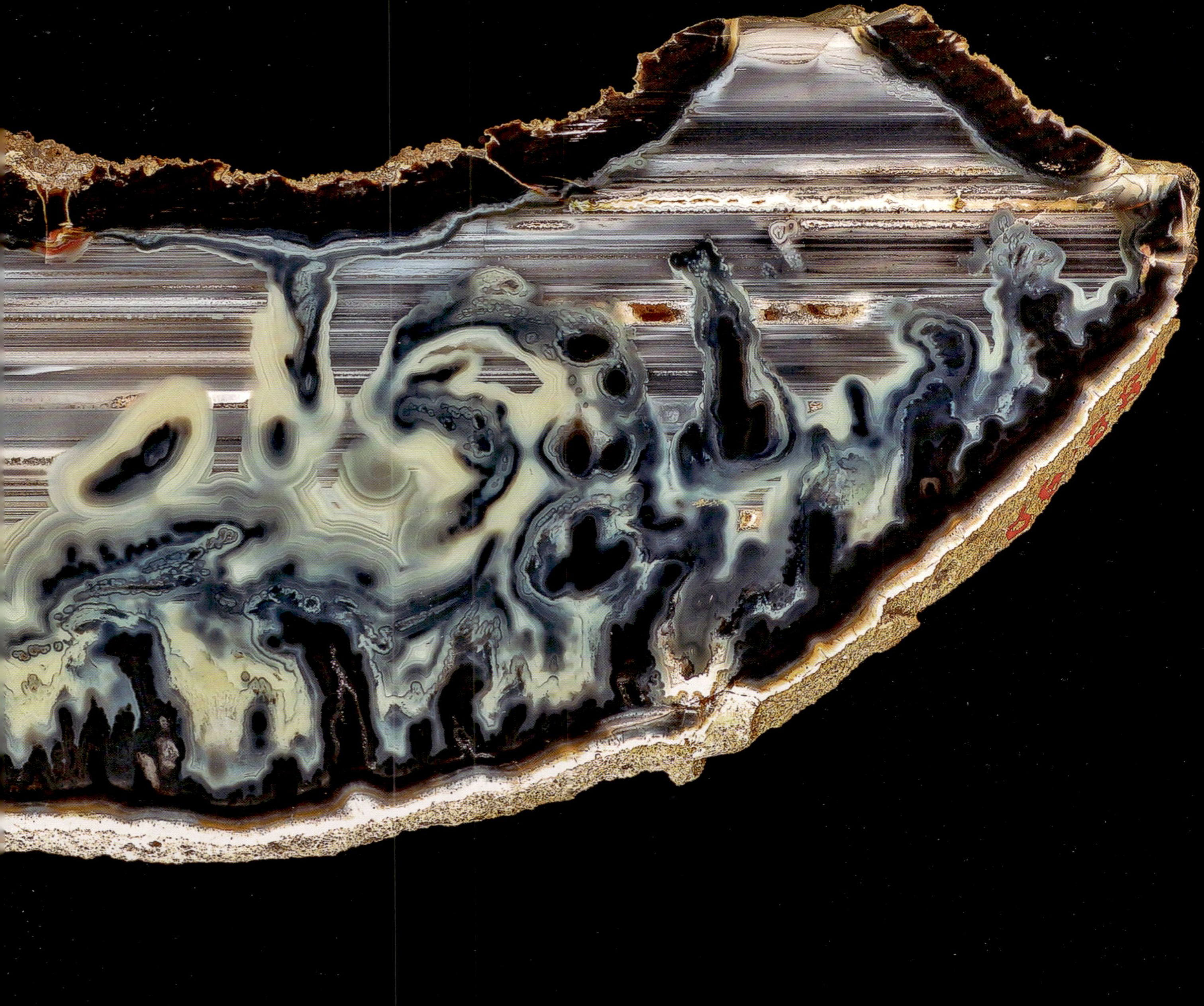

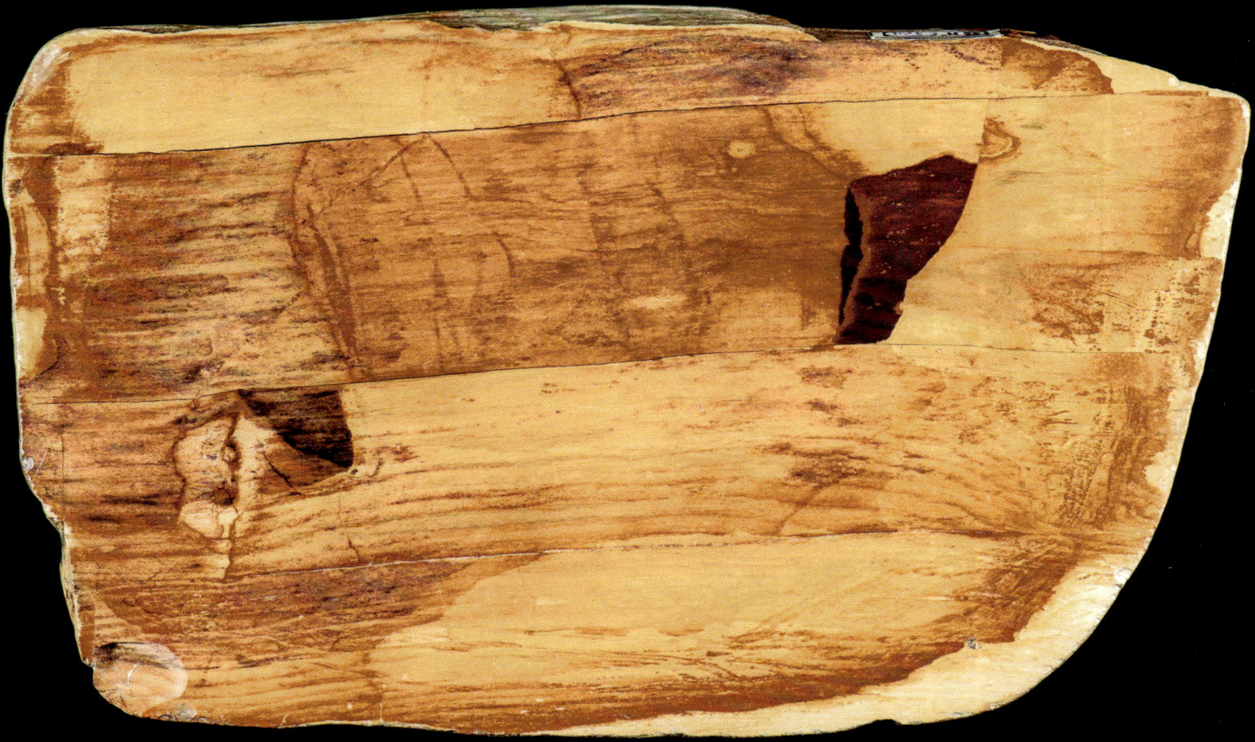

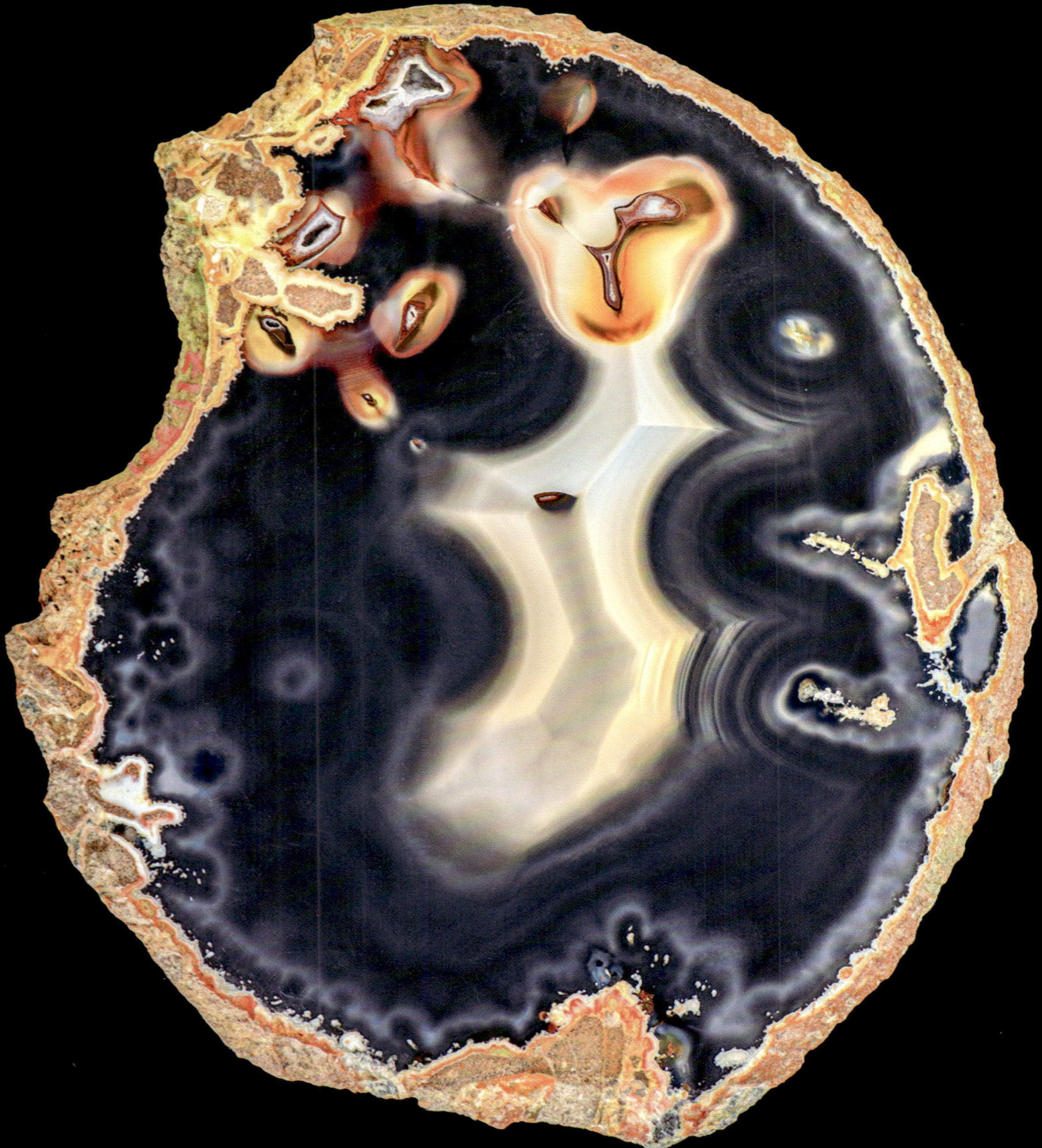

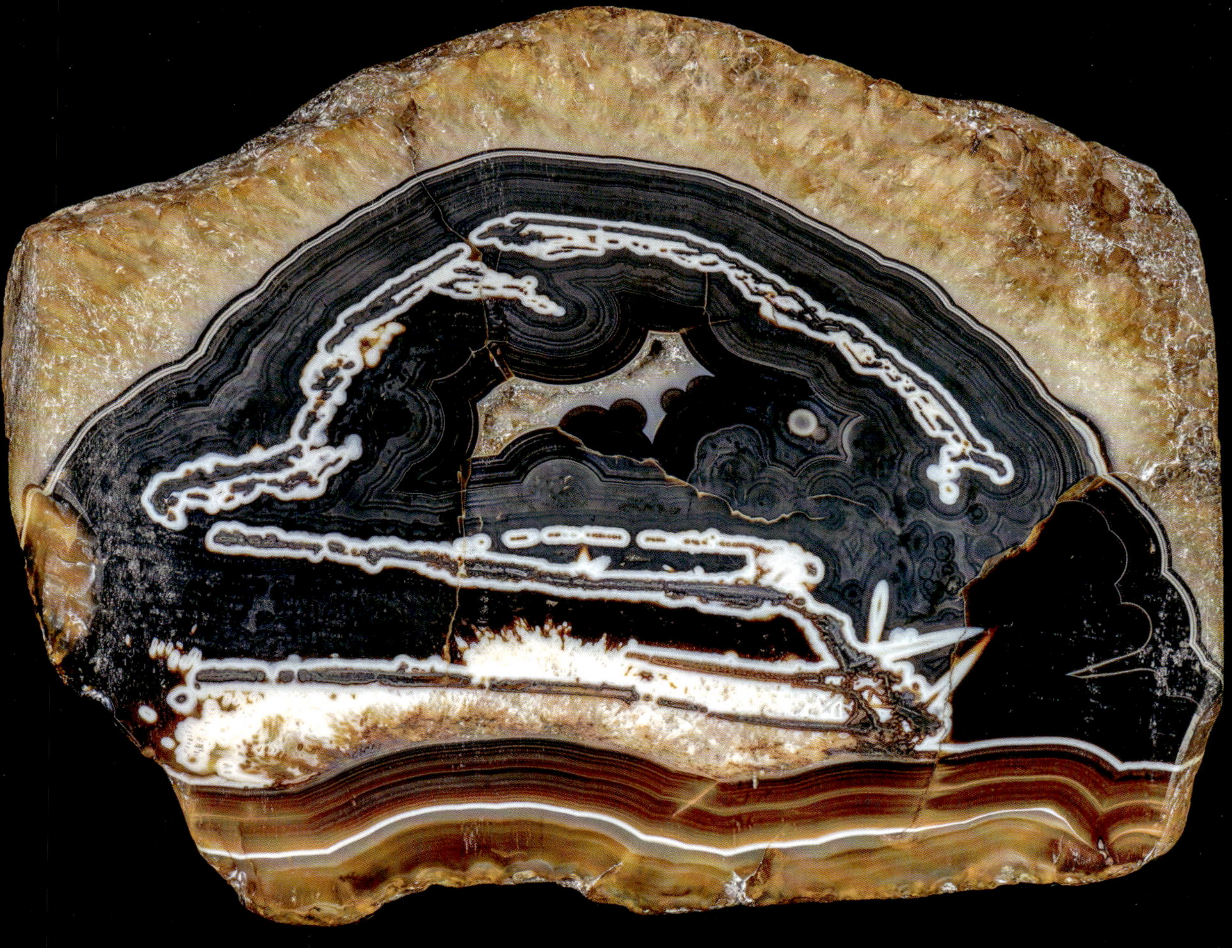

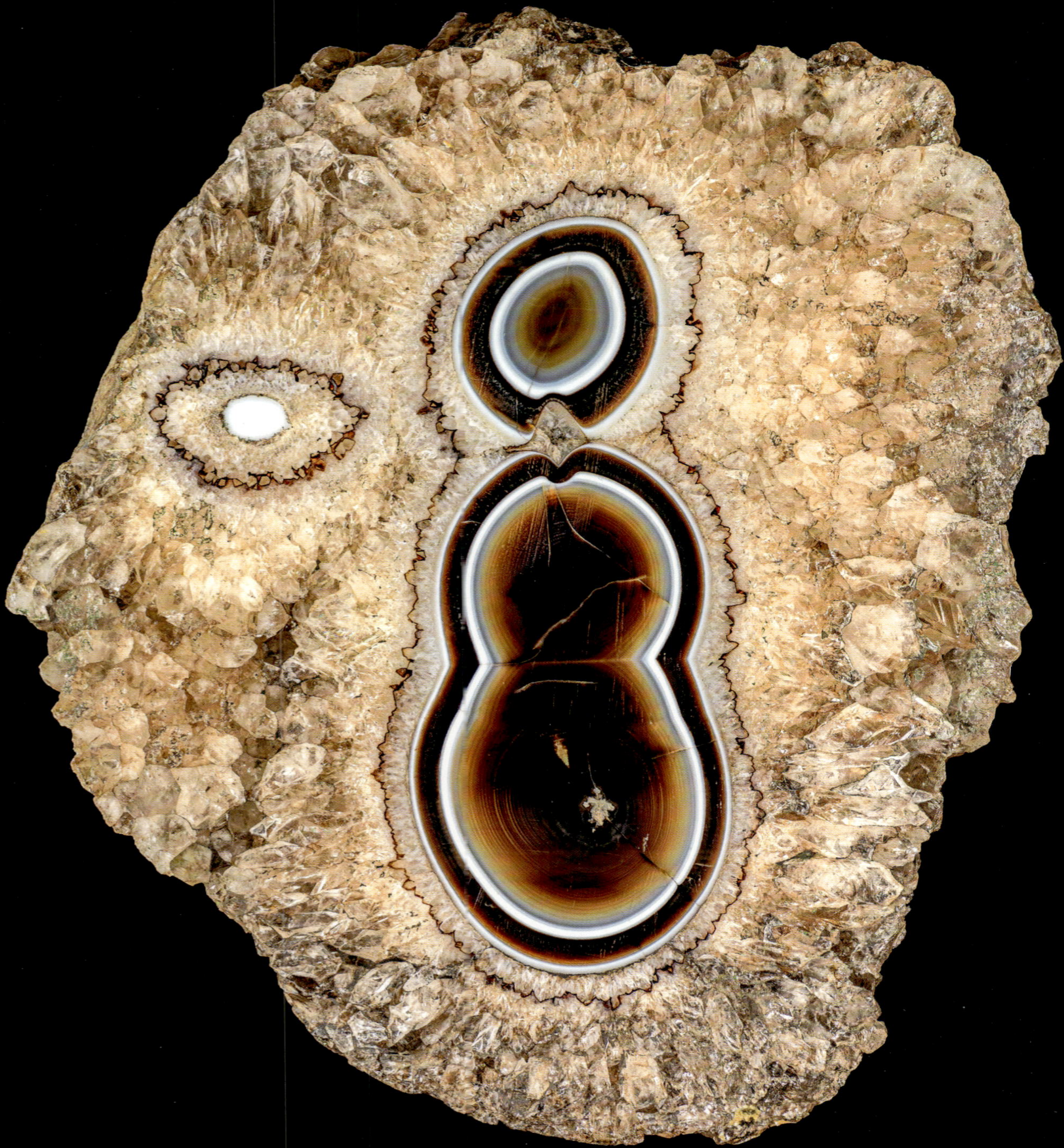

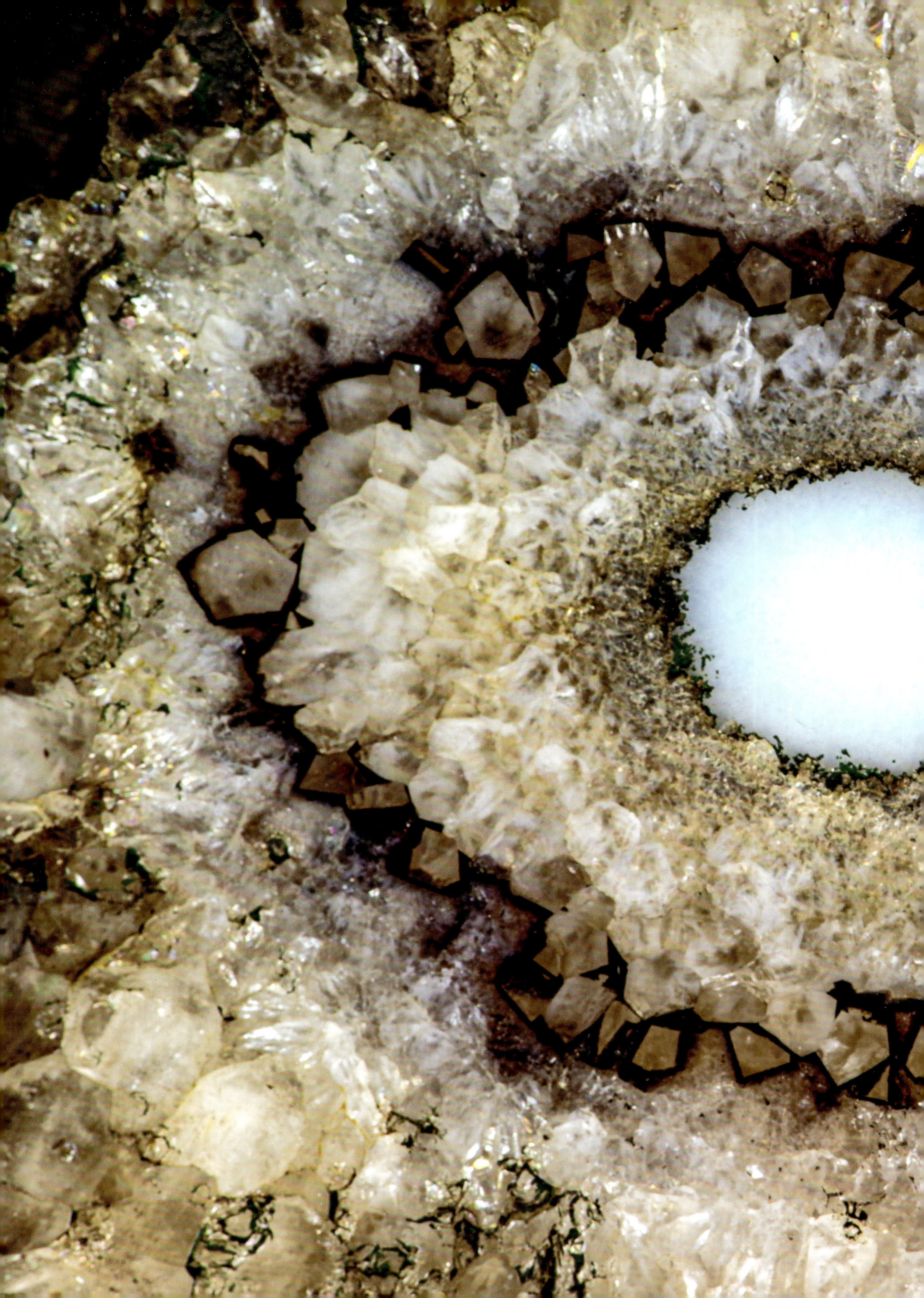

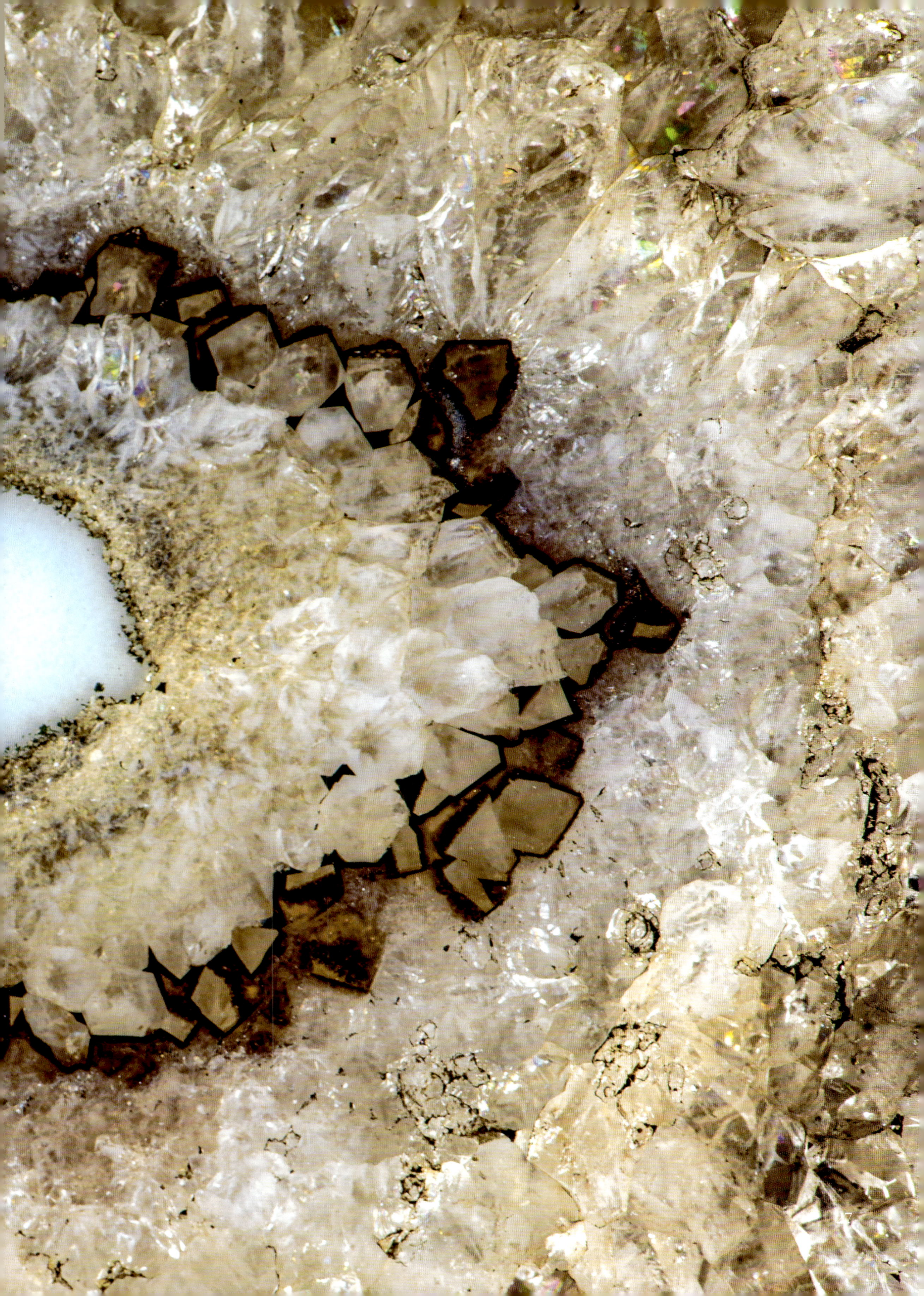

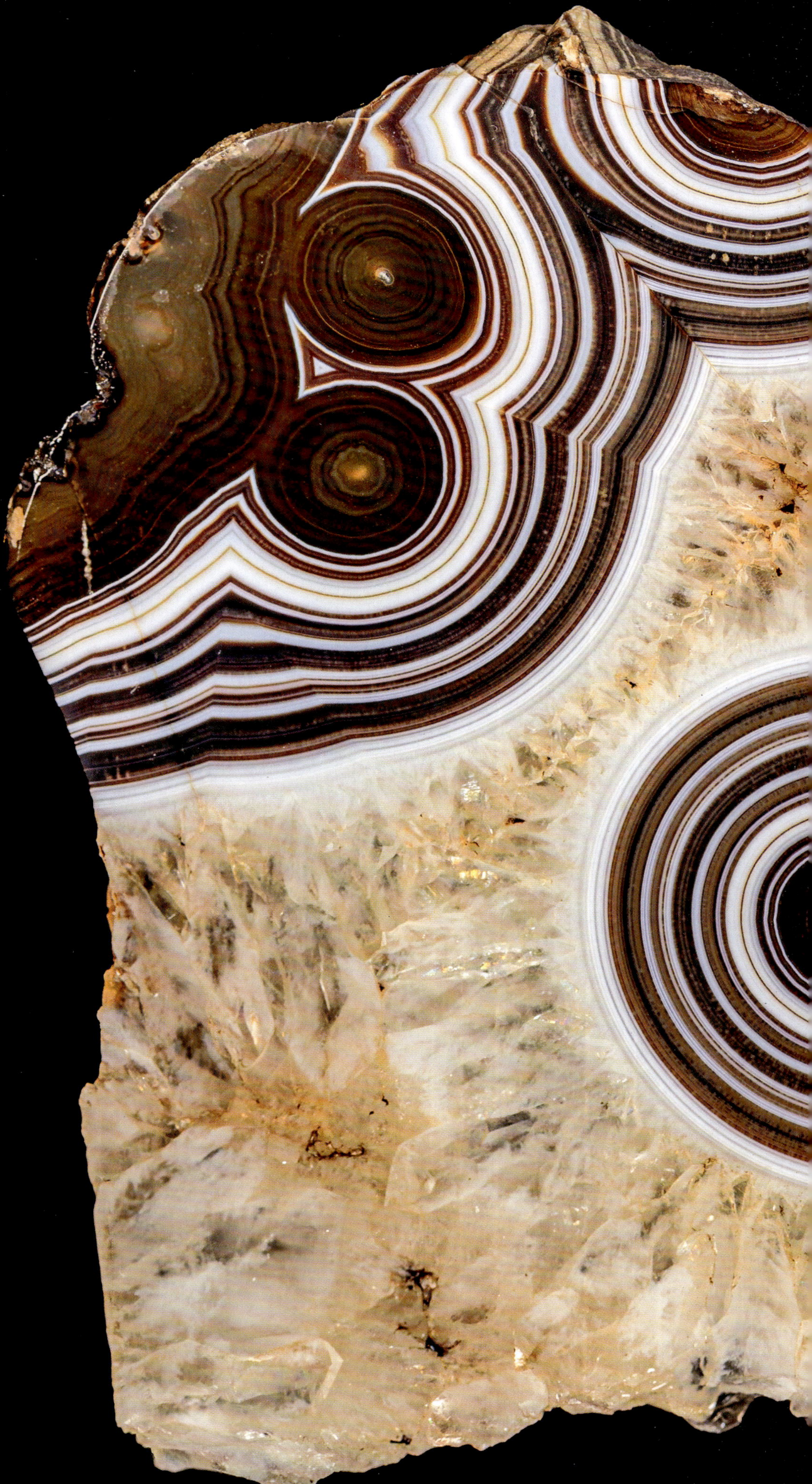

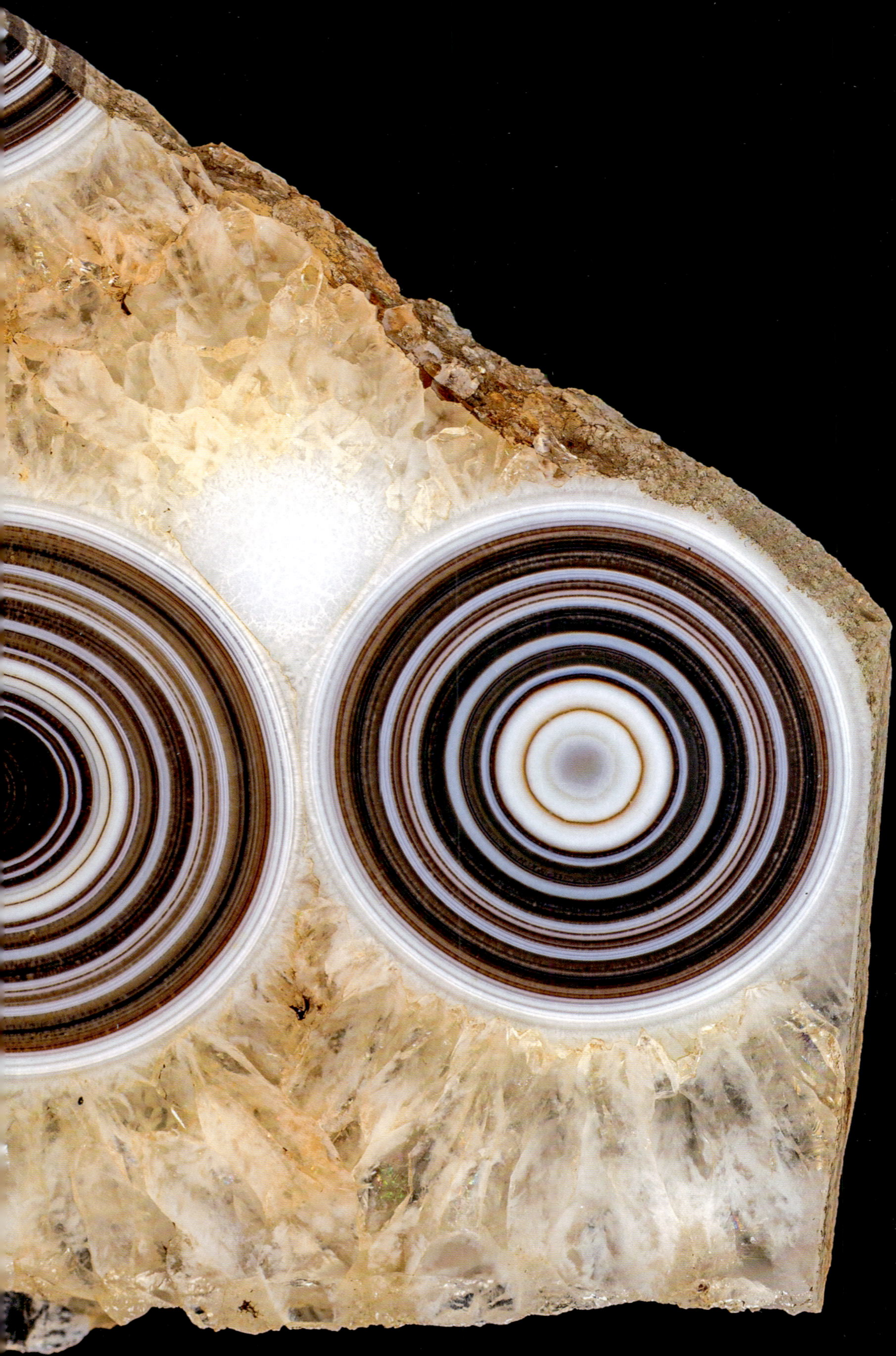

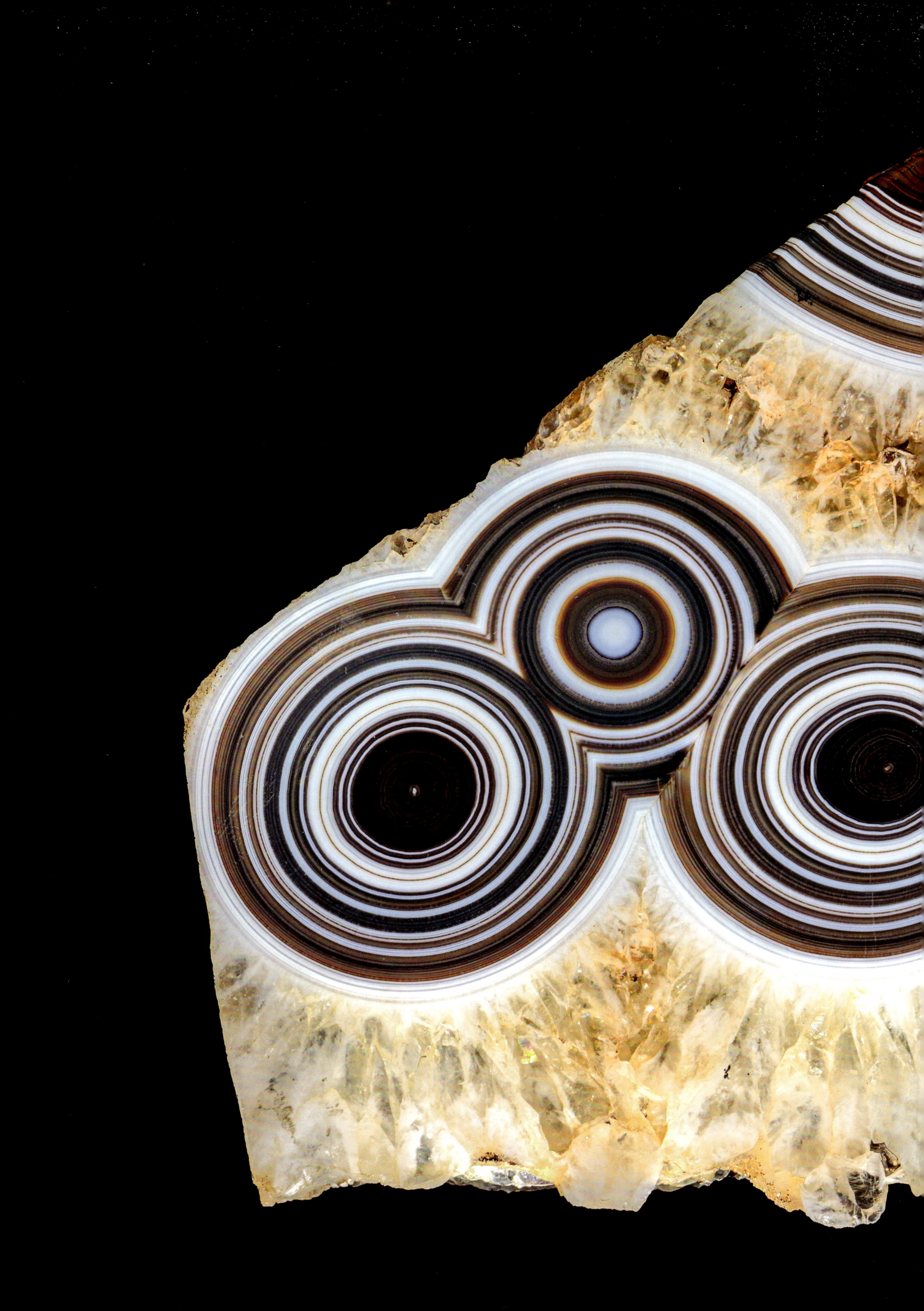

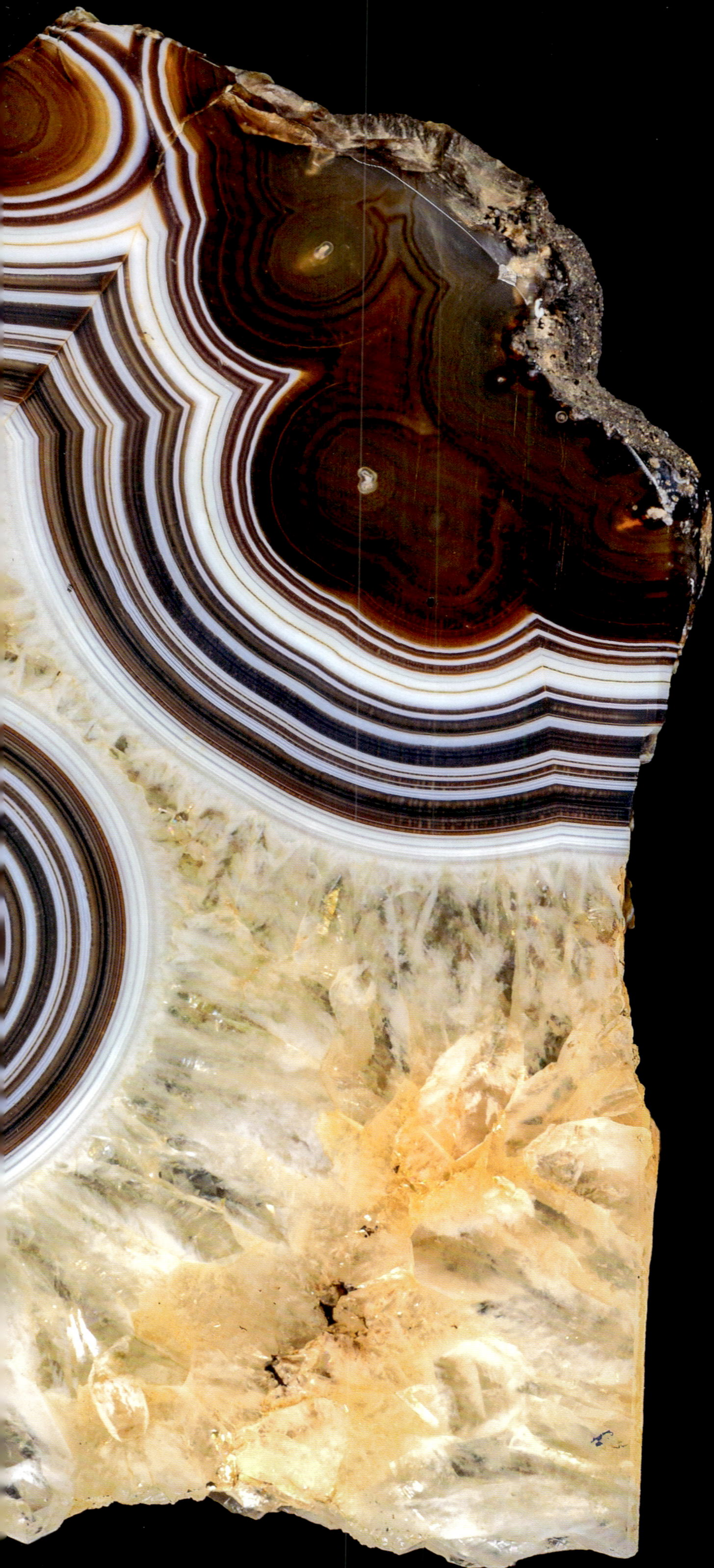

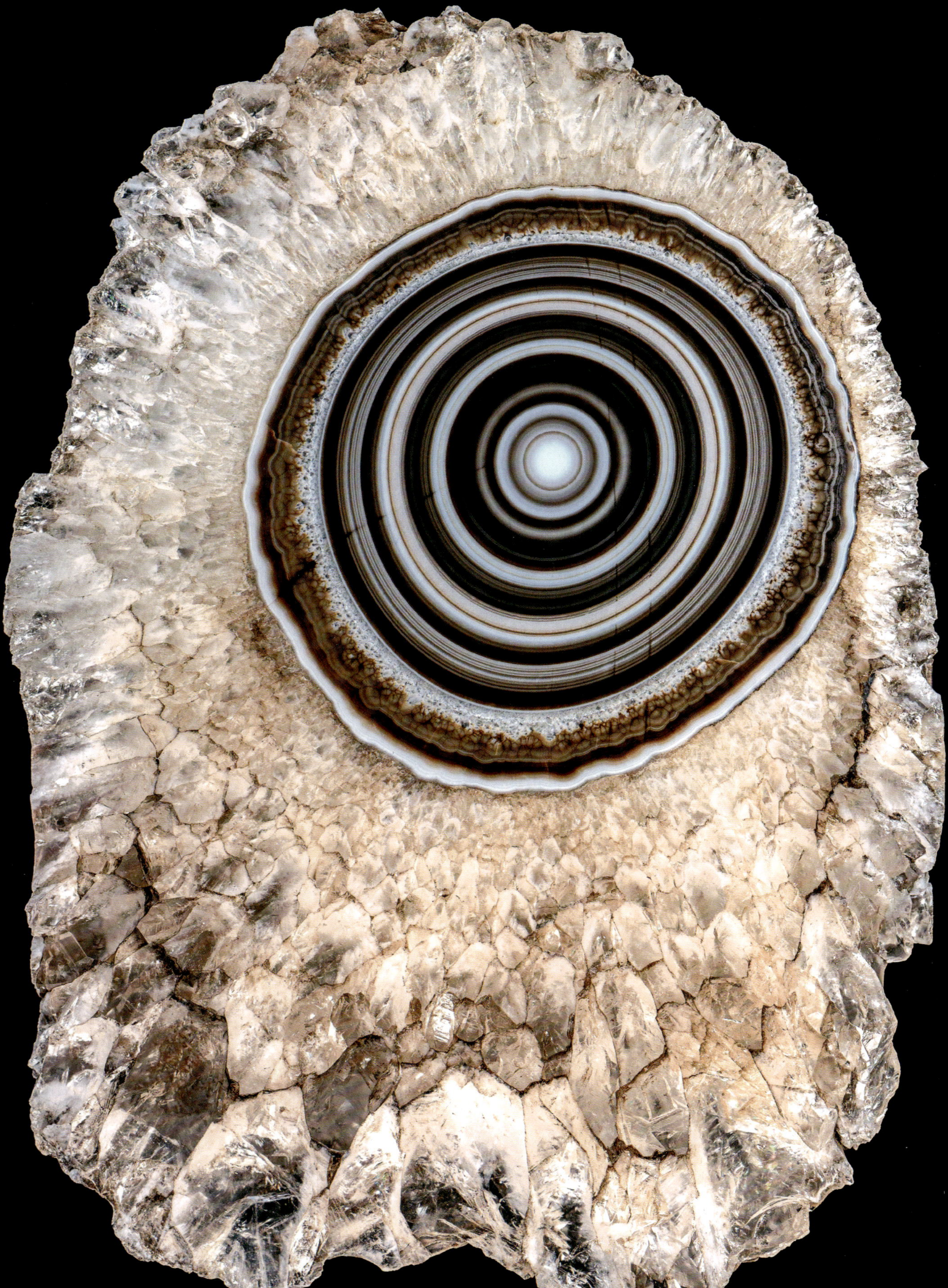

Mineral Ecstasy: Roger Caillois Writer and Reader of Stone

by Carlo Ossola

Desert Rose and Primeval Agate. Roger Caillois' "fantastique naturel"[1]

"Digito enim Dei Lex scripta est; sed propter duros in lapide scripta est" (*Exod.*, XXXI, 18): "For the law was written with the finger of God; but written on stone because of the hard-hearted," Saint Augustine observed.[2] Since the Old Testament, stone has been an early sign of unruly hardness and invincible foundation: "Thou art Peter, and upon this rock I will build my church" (*Matth.*, XVI, 18). In a formless world such as ours, it is helpful to come back to this and find – intact – the forms that have been deposited by the ages.

Roger Caillois, who believed in the "universal coherence"[3] of creation, revealed imperceptible worlds to us, in stones, at least as many as Athanasius Kircher had unearthed during the seventeenth century in his *Mundus subterraneus, in XII libros digestus* of 1665. Stone's mythology has been less rich than that of the vegetable kingdom, which has often – as demonstrated by Claude Lévi-Strauss – been taken as a model for the human species.[4] Moreover, in Ludvig Holberg's *A Journey to the World Underground by Nicolas Klimius*, 1741, what Klimius finds at the bottom of his descent is a forest of trees, more civilised and wiser than the human species that inhabits the earth's surface. Despite paying more attention than other writers to the new sciences, Jules Verne, in his *Journey to the Centre of the Earth*[5] – in which the three protagonists ascend from the extinct bowels of an Icelandic volcano to Stromboli, visiting numerous layers of the earth's crust – also names very few stone species, attracted, if anything, by the boiling waters and forests of gigantic prehistoric fungi. In the twentieth century, not even with Italo Calvino, who grappled with *The Stone Sky*[6] and contemplated more than one *Collection of Sand*,[7] did the narratives of this stone world, of which the globe that surrounds and supports us is made, truly increase in number. Even Francis Ponge, in turn deployed for the *Parti pris des choses* and the object-based *Pièces*, essentially left little room for the mineral world, limiting himself to: *Le galet*[8] and *L'anthracite.*[9] And finally Yves Bonnefoy, who dedicated many poems to stones, also delivers us to their sealed hardness, their obstructing presence:

Comme au vœu la pierre est fermée.
Que cherchions-nous ? Rien peut-être,
Une passion n'est qu'un rêve,
Ses mains ne demandent pas.[10]

However, in the century of positive sciences, alongside the disastrous experiences of the mines recounted by Johann Peter Hebel and later by Giovanni Verga,[11] there have also been moments of dizzying inventive tension. Moments when the world of stone asserted itself, initially with a nomenclature that was lingeringly analogical, as displayed by Adolfo Casali, with a contented desire for perfection and saturation of what is possible:

Aerophane stone.
Variety of Silica opal [...]
Alliance stone.
It is a variety of Syenite [...]
Amazons Stone.
It is a variety of Ortose [...]
Apocalypse Stone.
[...] variety of Resinite.
Eagle stone. The cavity of the geodes is sometimes filled with dusty matter; as it dries out, it often withers in a way that separates it from the walls of the cavity, within which it can move. [...]
Silver stone. Litargirio is known by this name [...]
Arithmetic stone. It is a variety of Pegmatite, which displays broken lines on its surface that simulate Arabic numerals.[12]

And so on, from *Aromatic stone* to *Asparagus stone*, from *Atramentaria*

Minera Cryſtalli, ex Muſeo Anidori.

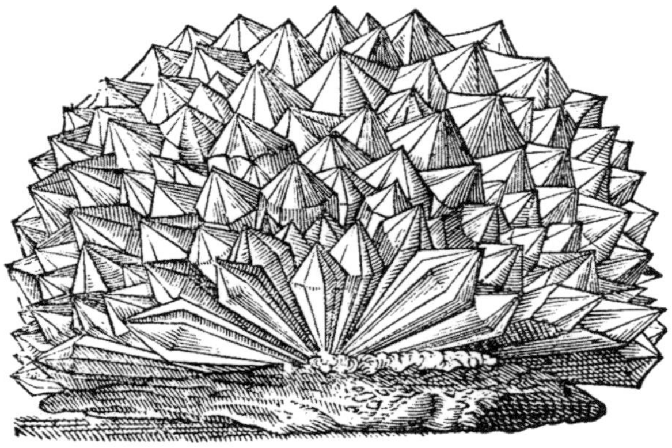

Pſeudocryſtallus.

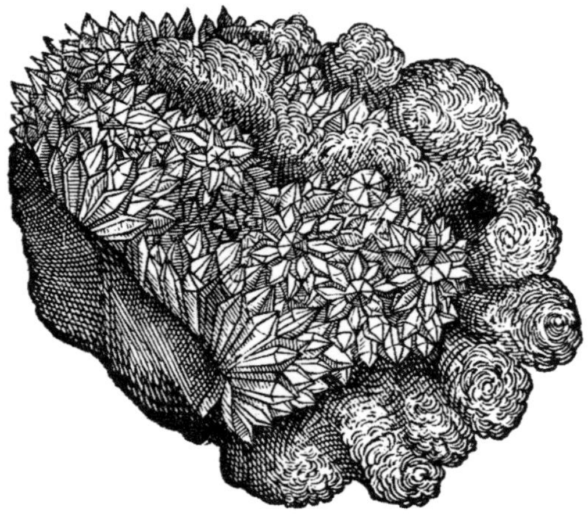

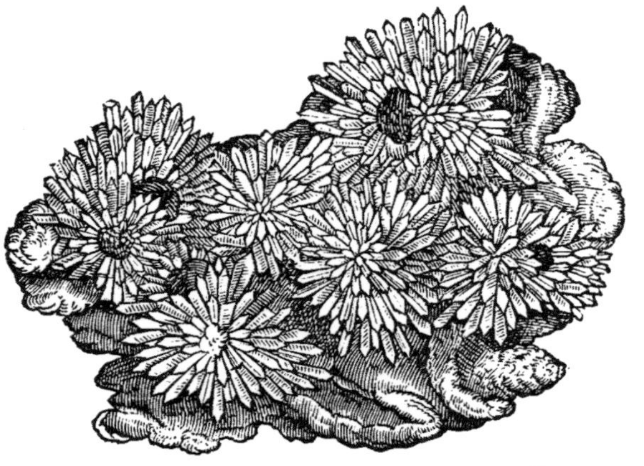

Formæ Topaziorum, Amethyſtorum, Beryllorum.

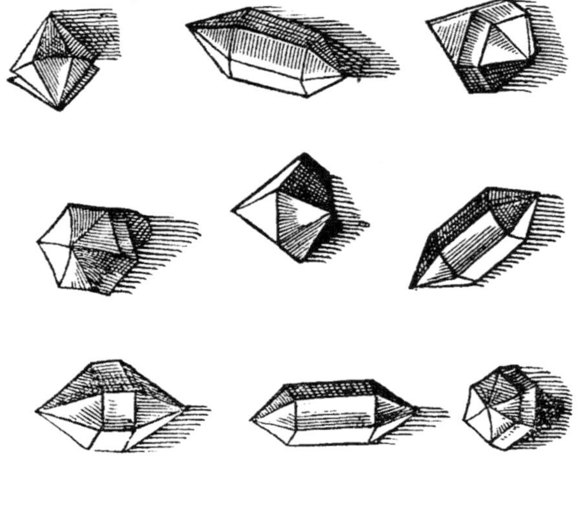

stone to *Aventurine stone*, from *Colubrine stone* to *Elastic stone* and *Mushroom stone*; extending the universal analogy over a succession of pages to *Tripe stone* and *Flying stone*.[13]

In the history of *curiositas* about "naturally painted" stones, an important role is given to Gaffarel, who, in his wonderful repertoire, *Curiositez inouyez sur la sculpture talismanique des Persans*, dedicates an entire chapter to the virtues of the "camaïeu," specifying:

But whether we want to admit fortuitous figures, we cannot fail to show the power of a great number that can only be called natural [...] GAMAHÉ: a word taken, in my judgement from Camaïeu, as figured agates are known in France, so that from a specific word a general one is made, suitable for all kinds of figured stones.[14]

This was followed by Carl Nicolaus Lang's *Historia lapidum figuratorum Helvetiae eiusque viciniae*,[15] richly illustrated and accompanied by fascinating plates. In the eighteenth century, fossils were, of course, also included in "lapides figurati;" it was not until the end of the century that a detailed classification would be possible in the *Systema Mineralogicum quo Corpora Mineralia in Classes, Ordines, Genera et Species, suis cum varietatibus divisa, describuntur* by Johan Gotskalk Waller.[16] And this success also determined its own "archaeology," as confirmed in good time by Pierre Barrère's *Observations sur l'origine et la formation des pierres figurées, et sur celles qui, tant extérieurement qu'intérieurement, ont une figure réguliere et déterminée.*[17] The author, in the dedication to his dissertation, already seems to anticipate the assumptions of the analogical method that permits (and would allow Caillois) to transition between the kingdoms: "Evidence is not seized upon everywhere, as you know Monsieur; most often possibilities are discerned and conjectures supported by analogy."[18] His definition includes the *Glossopètres*, which could be taken as ancestors and presuppositions of the *Writing of Stones*:

A figured stone, strictly speaking, is any stone that expresses exactly, when flat, hollow, or in relief, the features of various organised bodies. Such as Echinites, Glossopetrae, Toadstones, Dendrophores *and many others.*[19]

This tradition, in the Italian context,[20] is emphasised by the precious figured nature of "pietre paesine,"objects coveted by Baroque collectors and about which Jurgis Baltrušaitis wrote some unforgettable pages, *Aberrations. Quatre essais sur la légende des formes* in the chapter on *Pierres imagées*, dwelling at length, in turn, on Kircher and the symbolic functions of the *Gamahé*.[21]

But Caillois brought figured stones out from the realm of conceptual rarities of the mineral kingdom to that of the language of the universe, where – Lovejoy wrote – *The Great Chain of Being*[22] stretches, where forms imitate and contaminate each other, where the isomorphism of contiguity prevails over heteroclitic dissimiliarity. This was recognised by Marguerite Yourcenar in 1981 in her *Discours de réception à l'Académie française*, in which she celebrates – as is costumary – her predecessor Roger Caillois:

Far from disparaging what is human, as has been alleged, he [Caillois] found it all along a scale ranging from molecules to the stars. Because he claimed to observe the presence throughout the universe of a sensibility and a consciousness both analogous to our own, he has been accused of anthropomorphism. But Caillois himself has argued passionately that on the contrary he advocated an inverted anthropomorphism in which man, instead of attributing his own emotions, sometimes condescendingly, to all other living beings, shares humbly, yet perhaps also with pride, in everything contained or innate in all three realms.[23]

Previous page Athanasius Kircher (1602-1680), Mundus subterraneus in XII libro digestos, quo universae denique naturae divitiae, *1678. The image is taken from volume II. It reproduces crystals'geometrical shapes.*

He was the demiurge of the four elements, of their mutual "mimicry", always tending to seek hidden symmetries or Brunian "vinculis"[24]: "J'y distingue, sinon une réponse de la nature, au moins une des métaphores possibles de la *cohérence aventureuse* où j'aperçois pour ma part l'ambition commune de la science et de la poésie."[25]
His thesis is explicitly stated in his "research into the origin of the beauty of stones" in the preamble that justifies the *Morphologie générale des minéraux:*

Ces formes sont d'avant histoire, d'immémoriale seigneurie. En de telles structures, façonnées aux plus rudes traitements et ennoblies par eux, la loi d'équilibre l'emporte à la fin. Il en devait être *ainsi, immanquablement. Au commencement, au plus ardent du chaos, l'équilibre qui allait parvenir à tant de délicatesses miraculeuses ne fut sans doute rien d'autre que le jeu des compensations encore instables et grossières, qui, lentement, mettait fin aux soubresauts d'un astre en train de se figer. Peut-être n'est-il pas de plus sûrs modèles de la beauté profonde que les formes* émergées *des grandes acrimonies.*[26]

There is no doubt that this ambition, while it underlines the final harmony towards which the universe tends in the balance of compensations, is, on the other hand, fascinated by the possibility of making stone – in its perenniality – absolute, significant and signifying of itself:

To decipher such writing, if writing it is, would not mean trying to unravel an inextricable mass of lines and loops, but rather endeavouring to interpret anew some oft-repeated signs so turned in upon themselves that they refer only to their own form.[27]

Caillois' dream aims to fundamentally overturn the myth of Deucalion and Phyrrha, according to which, after the flood, these two survivors bring back the human race by throwing stones behind them that become living human beings. Caillois contemplates the opposite, that man, after millennia of presence on earth, looks at the gigantic stones raised up by his distant ancestors and in them contemplates homo erectus and the boundless anxiety at surpassing himself that has led to the present day:

les siècles ont passé. Les descendants lointains des constructeurs démunis disposent d'un pouvoir quasi illimité. Ils peuvent lire dans ces pierres, rien que des pierres mises à l'aplomb du sol [...], le premier témoignage d'une ambition obscure, la leur, aussi démesurée, aussi mal dégrossie et aussi solitaires qu'elles. Et ils admirent que des stèles difformes inaugurent l'histoire entière de leur espèce.[28]

Beneath Caillois' virtuoso writing, particularly elegant and refined in the *Writing of Stones*, circulates a vision that is both cosmogonic (on the differentiation of the elements emerging from the primordial chaos) and apocalyptic (on the ultimate destiny of created forms). The "morphological anxiety" that runs through Caillois' writing rests on one of the rarest and most precious principles – and one that has disappeared from our horizon today – *astonishment*, which nature, if delicately observed, arouses in us in abundance:

Nature itself provides [reasons for wonder] from time to time: an insect, a root, a fragment of lava, a fossil imprint in the anthracite of a giant fern, the rose-shaped desert stone, every wreck, finally, which has suffered the violence of the winds over the course of the centuries, the pressure of the great seabed or the action of high temperatures and which, having escaped these terrible laboratories, comes to land as if by miracle on our quiet shores, formed and deformed (we do not know which) by the primordial forces.[29]

"Desert rose" (alongside agate) was undoubtedly the mineral formation of which Caillois was most fond; the

Next page
Carl Nicolaus Lang (1670-1741), Historia lapidum figuratorum Helvetiae, ejusque viciniae, *1708.*

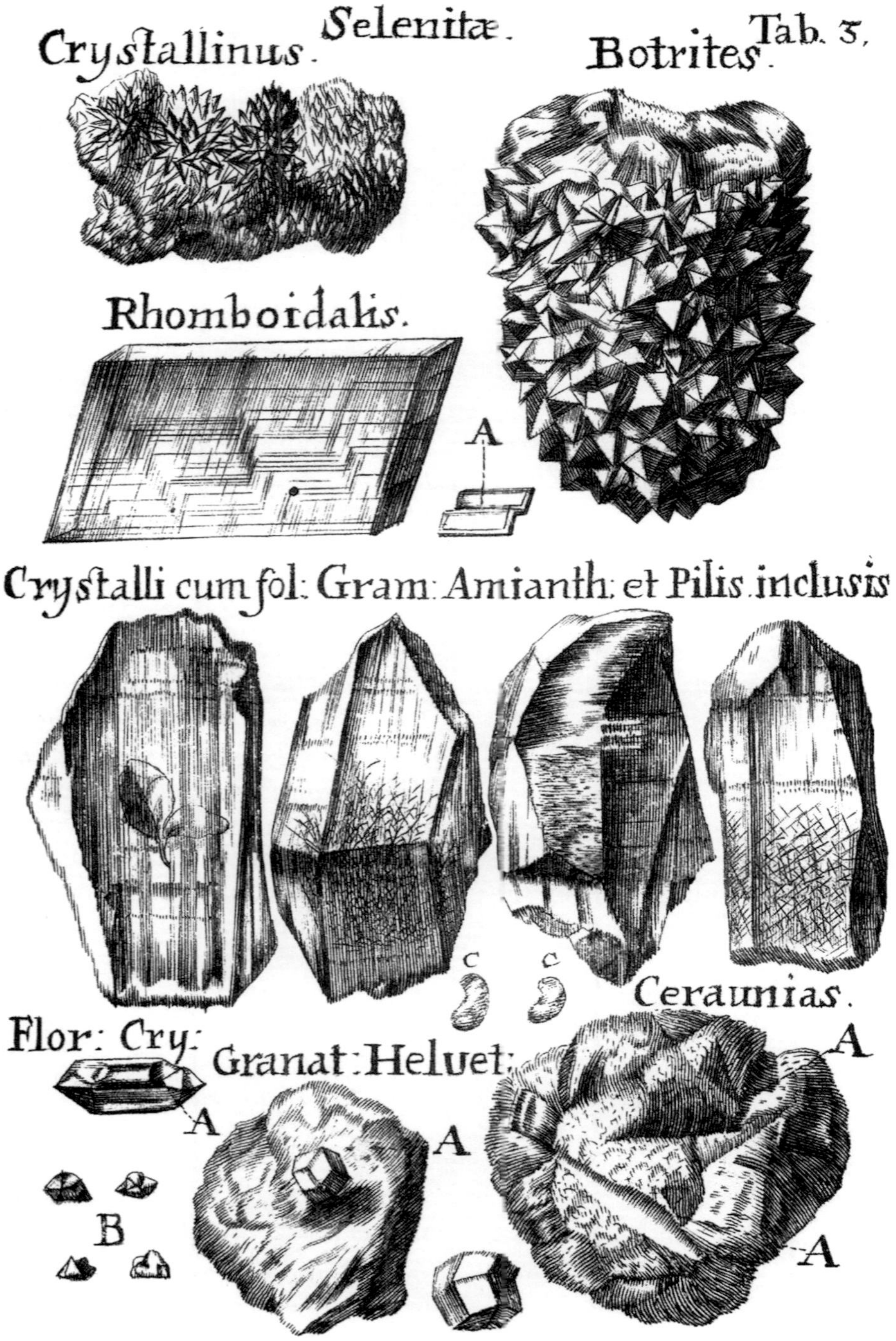
Crystallinus.
Selenitæ.
Botrites.
Tab. 3.
Rhomboidalis.
A
Crystalli cum fol: Gram: Amianth: et Pilis inclusis
C
C
Ceraunias.
Flor: Cry:
Granat: Heluet:
A
A
A
B
A

stone draws its stability, fascination and infertile yet splendidly evident immortality from its very aridity:

D'où vient, j'imagine, que les roses des sables sont appelées roses, quand elles sont le contraire d'une fleur, serait-elle de pierre. C'est peu qu'elles aient renoncé à toute souplesse et même au ruissellement entre les doigts des grains brillants qui les composent et qui, du moins, avaient comme l'eau la propriété de couler. L'imagination commence ici ses fables, ses délires. Elle voit en ces objets revêches et durs un hymne superflu du désert à la sécheresse ; ils paraissent faire don à l'astre qui évapore *d'une plus dense offrande de stérilité. Je sais: quoi donc est immortel qui d'abord n'est pas stérile?*[30]

This passage succinctly reveals the two traces in Caillois' work that directed, as symbol and recapitulation, his thirst for eternal plenitude, unchanged and no longer metamorphic beyond that culmination: stone and water, as described in this exemplary apologue. I will allow myself only to suggest one example for each of these paths indicated by Caillois: on the one hand, the concentration of something that cannot be changed in the figured stones, which the concentric rings of the agate fix as motionless solar systems:

En même temps, ces armoiries sont norme et canon de la beauté profonde, celle que, sur le rivage opposé, les rares réussites du génie s'efforcent d'enrichir ou de retrouver. Elles procurent en outre, prise sur le vif et à tel instant de son progrès, une coupe irrécusable faite dans le tissu de l'univers. Comme l'empreinte fossile, ce sceau, cette trace n'est pas effigie seulement, mais la chose elle-même par miracle stabilisée, qui témoigne de soi et des lois cachées de la lancée commune où la nature entière est entraînée.[31]

This "irrefutable" exemplarity, mould and model of the laws of harmony that regulate the universe, enclosed in a stone, is symmetrical of the other "harmony" that comes from the fusion of the elements into a single flowing of water, foams, vapours and irises. It is the myth, mentioned above, of water that descends or ascends to an origin which is also its own end: river water that mixes with the ocean, or a fluid form of Arethusa turned into a wave that joins with other waves in the river of Alpheus, it too now indistinct flowing water, according to Ovid's story. But what counts here is that Caillois made this myth his own emblem, the very symbol of returning to the origin, endlessly losing one's way:

Fleuve rescapé du naufrage, je séparai mes eaux, je les ressemblai, je leur creusai un estuaire, qui était *un nouveau début. Il me fallut en ce point altérer la légende. Fleuve issu de la mer, le fleuve Alphée ne saurait* être *un fleuve comme les autres, mais un fleuve inverse et pour ainsi dire symétrique. Je l'imagine à bout d'élan et de forces, remontant les pentes, coulant à rebours, comme un film qu'on déroule à l'envers. Son débit s'amenuise à mesure. En revanche, il gagne en limpidité. Il se trouve heureux de s'approcher de la fissure où il disparaîtra et qu'il devine déjà semblable à celle qui lui a donné naissance avant son* équipée *maritime, modeste, insignifiante, comme sont les sources véritables qui laissent couler les fleuves vers leur embouchure, leur delta ou, comme il arrive aussi, qui les abandonnent et les oublient. Alors, ils ne sont même pas la proie de l'*étendue *infertile. Ils sont bus par les sables des déserts ou sont engloutis dans quelque perte mystérieuse, imprévisible.*[32]

In a perfect circle, we come back to the waters of the desert, to the "rose des sables" that seemed to Caillois like a "ruissellement entre les doigts des grains brillants qui les composent et qui, du moins, avaient comme l'eau la propriété de couler". It would, however, be a question of an, albeit wonderful, circularity and not of harmonious fusion: water and sand would return, alternatively, one in the other, purity of sterile immortality: "quoi donc est immortel qui d'abord n'est pas stérile?" Yet at least once Caillois succeeds in finding, one in the other, stone and water, in contemplating the quivering drop in an agate glare, in contemplating the perfection of the universe yearned for by a passing guest who has nothing of those immaculate origins if not regret:

Les deux faces de l'agate sont également *polies et du même bleu nocturne.* [...] *Entre elles, qui semble en garantir la terrible promesse, l'eau cachée des origines dont on voit l'ombre se déplacer et dont l'oreille entend le clapotis. Je crois que nul ne reste insensible à l'*émotion *qu'engendre pareille présence.* [...]
Le vivant qui le [ce fluide incorruptible] regarde comprend qu'il n'est, pour sa part, ni si durable ni si ferme. Ni si agile ni si pur. Il se connaît sans joie à l'extrémité d'un autre empire, et soudain si étranger *à l'univers : un intrus hébété. Je ne devine que trop, par obsession personnelle, quelles méditations, du moins quelle rêveries vagues, un passager du monde peut commencer à dévider à partir de ces cailloux hantés d'une liqueur, un peu d'eau géologique restée prisonnière dans la poche transparente d'une pierre hermétique.*[33]

This is why reading Caillois, and contemplating his "obsessions," is always the gift of skimming our origins and accessing the immutable.

Notes

1. *The definition is by Roger Caillois himself,* L'agate de Pyrrhus, *in* Images, images..., *Paris, José Corti, 1966; all quotations will be taken from: R. Caillois,* Œuvres, *Paris, Gallimard [«Quarto»], 2008; this quote is from p. 733.*

2. *Aurelio Agostino,* In Joannis evangelium tractatus CXXIV*:* TRACTATUS XXXIII: *«Ex illa ergo turba cum audissent hos sermones ejus, Nec ego te condemnabo; vade, et amplius noli peccare»* (PL, *XXXV, 1649).*

3. *R. Caillois,* De la féerie à la science fiction : l'image fantastique, *Chapter 1 of* Images, images...; *ed. cit., p.679.*

4. *«Le végétal est le modèle de l'homme»: cited from Pausania [VIII, XXIX, 4] mentioned and commented on by Claude Lévi-Strauss in* Anthropologie structurale, *Paris, Plon, 1958 and 1974, p. 239.*

5. *Paris, Hetzel, 1864.6. I. Calvino,* Il cielo di pietra, *in* La memoria del mondo e altre storie cosmicomiche, *Turin, Einaudi, 1968 and 1975, pp. 78-84.*

7. *Id.,* Collezione di sabbia, *1974; story from the volume of the same name* Collezione di sabbia, *Milan, Garzanti, 1984, pp. 9-13.*

8. *F. Ponge,* Le parti pris des choses, *1942: later in* Œuvres complètes, *Paris, Gallimard, 1999, vol. I, p. 49.*

9. *Id.,* Pièces, *1961; later in* Œuvres complètes, *cit., vol. I, pp. 732-733. With the addition of the two pages* Des cristaux naturels *(*ibid., *pp. 632-633) the inventory will be almost complete.*

10. *Y. Bonnefoy,* Une pierre, *from* La Vie errante, *1993; now in* Œuvres poétiques, *Paris, Gallimard, 2023, p. 682.*

11. *J. P. Hebel,* Insperato ritrovamento [Unverhoftes Wiedersehen], *in* Storie di calendario [Kalendargeschichten], *1811; cited from the bilingual edition edited by Giuseppe Bevilacqua, Venice, Marsilio, 1996, pp. 272-279. G. Verga,* Rosso Malpelo, *1878; later in* Novelle dei campi, *Milan, Treves, 1880.*

12. *A. Casali,* Dizionario delle denominazioni e dei sinonimi della chimica e delle scienze arti e industrie attinenti alla medesima, *Bologna, Nicola Zanichelli, 1872, p. 330*

13. Ibid., *pp. 331 – 336.*

14. *J. Gaffarel,* Curiositez inouyez sur la sculpture talismanique des Persans, *À Paris, chez Hervé du Mesnil, MDCXXIX, Chap.V:*

Preuve de la puissance des images artificielles par les naturelles, empreintes aux pierres et aux plantes, appelées vulgairement GAMAHÉ ou CAMAÏEU, ou Signatures, *pp. 154-155. Gaffarel also recalls* L'agate de Phyrrus, *studied by Caillois (cfr. here note 2), on pp. 156-157.*

15. Venetiis, Typis Jacobi Tomasini - Lucernae, apud Haeredes Gottofredi Hautt, MDCCVIII. The eighteenth century, with its inextinguishable vocation for taxonomy and encyclopaedia, also saw the creation of the catalogue by Johann Wilhelm Baumer, Historia naturalis lapidum pretiosorum omnium: nec non terrarum et lapidum hactenus in usum medicum vocatorum, [...], *Francofurti, apud Ioan. Georg. Fleischer, MDCCLXXI*

16. Viennae, ex officina Krausiana, 1778, vol. II: «in quo Minerae et Concreta describuntur», in particular § 162: Litoglyphi, Litomorphi, Lapides figurati. *The same author, in his* Brevis introductio in historiam litterariam mineralogicam atqve methodum *(Holmiae – Upsaliae, in Officinis M. Swederi, MDCCLXXIX. p. 73) still distinguished between* Fossilia concreta *and* Fossilia petrificata.

17. Paris, d'Houry père et L. d'Houry fils, 1746.

18. Ibid., *p. 1.*

19. Ibid., *pp. 1-2.*

20. See, at least, the volume by Roberta Coglitore, Pietre figurate. Forme del fantastico e mondo minerale, *Pisa, ETS, 2004.*

21. Paris, Olivier Perrin, 1957, Chap. II, pp. 47-72. Naturally Baltrušaitis's essay did not escape Mario Praz, who referred to it in I marmi di Roma, *in* Panopticon romano secondo, *Rome, Edizioni di Storia e Letteratura, 1977, p. 35.*

22. A. Lovejoy, The Great Chain of Being: a study of the history of an idea, *Cambridge (MA), Harvard University Press, 1936*

23. Cfr. p. 129

24. "The fictitious image entails its own truth" (Giordano Bruno, De vinculis in genere, *1591, (Roger Caillois' epigraph to the essay* Au cœur du fantastique, *Paris, Gallimard, 1965; now in* Œuvres, *cit., p. 837).*

25. R. Caillois, Agates paradoxales, *in* Cohérences aventureuses, Préface générale, *1976; and in* Œuvres, *cit., p. 809.*

26. Id., Pierres, *1966;* II: Physique, *Argument:* Morphologie générale des minéraux, *in* Œuvres, *cit., p. 1046.*

27. R. Caillois, L'écriture des pierres, Agates, *in* Œuvres, *cit., p. 1092; cfr. p. 104.*

28. Id., Pierres, IV: Morale, *in* Œuvres, *cit., p. 1074.*

29. Id., Vocabulaire esthétique, *Paris, Fontaine, 1946.*

30. Id., Rose des sables, *in* L'écriture des pierres*; in* Œuvres, *cit., pp. 1099-1100.*

31. Id., Pierres, V: Testament, Soleils inscrits, II; *in* Œuvres, *cit., p. 1083*

32. Id., Le fleuve Alphée, *1978; poi in* Œuvres, *cit.,* Prélude, *pp. 87-88.*

33. Id., Hors série. L'eau dans la pierre, *in* Pierres, II: Physique; *poi in* Œuvres, *cit., p. 1064.*

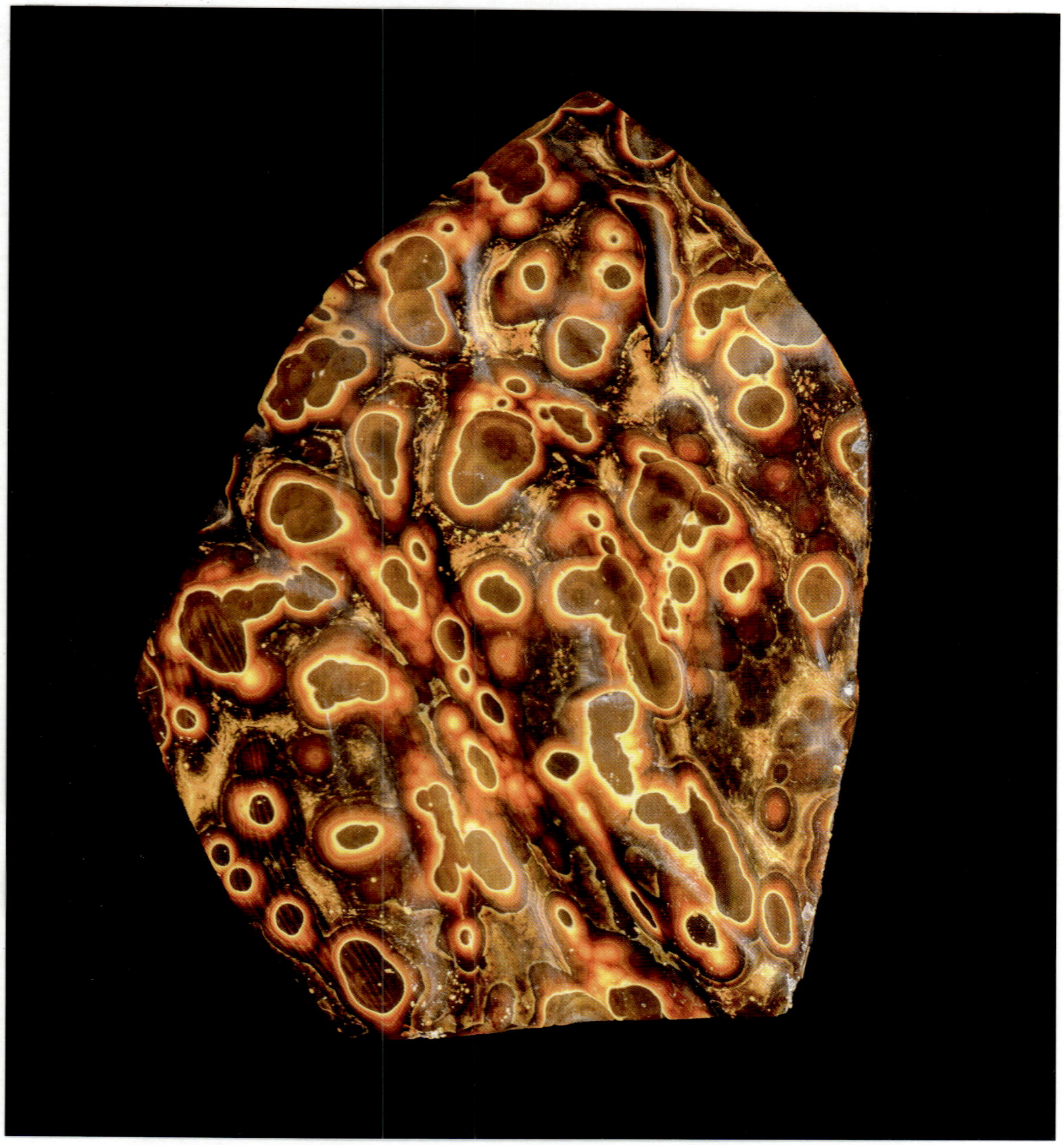

Roger Caillois's Estro Poetico*: of Minerals, Words, and Dreams*

by François Farges

Roger Caillois was born in 1913 in Reims as the walls were already beginning to crack, anticipating the roar of German cannons that would soon largely destroy this second capital of the Champagne region. What a cruel fate for this sacred land, whose cathedral had seen French kings crowned, to now face bombings and fires. As a young boy, Roger wandered through the city quite amazed, among the brightly coloured shards of stained-glass windows that the sun transformed into precious gems sparkling like rainbows while the rubble, with its smell of burnt stone, revealed curious remains, like the heads of medieval angels, charred but still smiling despite the scars left by cannon fire. Fire and Earth.

Among the ruins of the newfound peace, young Caillois was tireless, even painfully excited. Only water, lashing his face, could momentarily calm him. This bright young spark found an outlet in field and forest, breathing in the curious smells of nature, the scent of wheat and marcasite released from the ploughed earth by savage rains that cleansed the soul along with the air.

In the vicinity of the Aisne à la Marne canal, at number 75 rue de la Hincmar according to Caillois, although the actual address may rather have been either 37 or 39, the boy's mother tried desperately to get him to read, and the unruly seven-year-old confessed to being caught up in a "peaceful and ignorant" childhood. Air and Water.

All that remained was ether to unite these four pillars. Caillois befriended at the time one of his young classmates, Roger Gilbert-Leconte, who was brilliant in math and a talented writer. The boy, who would become a great poet – though he remains so little-known today – instilled in Roger certain foundational elements that triggered a literary breakthrough. At thirteen, he became one of the brightest in his class: but good marks and prizes were not enough for him. As a teen, slowly at first, then avidly, he discovered the "great authors" in order, he later confessed, to understand the literary minds of France. A place where, after all, the Curies, the Einsteins, and other scientific writers of the world, prolific as they were, did not count for much.

As a young man, in 1929, he "went up" to Paris where he met Yvette Billod – who would become his first wife – at the Lycée Louis-le-Grand which he attended in order to prepare for the entrance exams to the École normale supérieure ("rue Ulm" for those in the know) to which he was accepted in 1933. There he obtained his teaching diploma, frequenting André Breton and the surrealist circle for a few months, just enough to make their acquaintance. Caillois co-founded the journal *Inquisitions*, using the plural to avoid the ire of Breton who would not tolerate the overly qualitative singular term. And he immersed himself once more in the "surrationalism" espoused by Gaston Bachelard, the philosopher and epistemologist who was key to the connection between science and poetry that would lead Caillois, slowly but surely, to his "pierres de rêves" or "dream stones."

In the meantime, Roger Caillois founded the Collège de sociologie with Georges Bataille and Michel Leiris, where he met the Argentinian writer and philanthropist Victoria Ocampo in 1938. She invited him to stay in Argentina during the Second World War: a detour to Brazil allowed him to receive as a gift a so-called "ghost" quartz crystal and he was struck by the inclusions that seemed to "inhabit" the seemingly cold sample. This talisman was a missed opportunity, as it was confiscated by customs officers upon his return to Argentina. Caillois returned to his journals and founded another one with the Institut Français in Buenos Aires. He was involved with various journals at the time, including *Lettres françaises*, *La France libre*, *Confluences*, and *La Licorne*. After he returned to France, he worked with Gallimard on the collection *La Croix*

du Sud and on *Diogène*, a journal published by UNESCO, for whom he worked as an administrator (1948). After his divorce from Yvette, Caillois met Alena Vichrová, a political refugee from Czechoslovakia, whom he married in 1957. At the time, Caillois was laying the foundations of his "diagonal sciences" whereby he espoused connections to be found in the natural world. He discovered the metallic blue iridescence of a labradorite that reminded him of the morpho, an isochromatic butterfly native to Brazil. He bought it, and it proved fatal. Caillois' soul was slowly overtaken, and he caught the bug for collecting minerals, which opened his eyes to a little-known, crystalline world that was both eternal and fully formed.
During the same period, Caillois translated Borges into French and published numerous well-respected works, several of which were even published in other languages, mostly with Gallimard, the most notable of which are *Esthétique de Saint-John Perse* (1954), *Méduse et Cie* (1960), *Ponce Pilate : récit* (1961), *Jeux et Sport* (1967), *Cases d'un échiquier* (1970) and various texts on minerals including *Pierres* (1966), *L'écriture des pierres* (Skira, 1970), *Pierres réfléchies* (1975), and an autobiographical text, *Le fleuve Alphée* (1978).

Immortality

Roger Caillois was elected to the Académie française in 1971: the jeweller Jean Vendome, who created his ceremonial sword, would become one of his closest friends. His collection of insects, that had been given to the Muséum national d'histoire naturelle (at the Jardin des plantes in Paris), was unfortunately not kept together once his lavish purchases of minerals began to take off, particularly agates and paesines, two of the "dream stones" that were emblematic of his poetic explorations. Roger Caillois' health declined quickly as a result of various excesses: he had persistent high blood pressure, was averse to moderation, consistently gambled with his health, and was decidedly stubborn, and he died suddenly on December 21, 1978 at the age of only 65. He was buried in Montparnasse cemetery. Marguerite Yourcenar succeeded him at the Académie française after giving a eulogy to Caillois to its members entitled *L'homme qui aimait les pierres* (The Man Who Loved Stones). As if it was a love that could only be paradoxical, even unnatural.
Taking her place in chair number three under the Coupole des Immortels (The Dome of the Immortals, as members of the Académie française are called), Marguerite Yourcenar concluded the eulogy to her predecessor sublimely, noting in particular that *he had taken to its conclusion that "profound acquiescence" which, it may be assumed, he had already given while still alive. He no longer needed to question or to think; as one of Ionesco's characters had so aptly put it in* The King is Dying, *he no longer needed to breathe. The minerals of which he was composed belonged once more to the soil that had given birth to those beautiful objects he had never ceased to love.*
So much for the dietary constraints of a healthy life; they have deprived us of magnificent writing, perhaps even a Nobel Prize for literature, commentary on music that was 'nothing but noise' to his ears, (im)moral pronouncements on Léon Pierre-Quint's 'abnormality', pitiful arguments with Aimé Césaire, vain debates about the human and social sciences with Claude Lévi-Strauss, vitriolic attacks on Picasso and, more generally, on modern art and its champion, André Malraux, not to mention the conflation of mineral collecting with the science of minerals, mineralogy, which is not learned by collecting but by experiencing.

After Caillois' death, once the dust had settled and the page had been turned, relatives, friends, and various other hangers-on rushed to his widow Alena's side to get hold of unpublished manuscripts and spectacular samples, that is to say, those that were expensive and could be resold. Like André Goezu, Jean Vendome was outraged by this, or so I was told. Alena, in her generosity, gave away many pieces to mourners and other confidants of the time, some of which they later sold at a high price. In the meantime, between 1983 and 1985, she had finalised a donation to the Muséum national d'histoire naturelle, an undertaking wisely initiated by Henri-Jean Schubnel, then professor of mineralogy at the Jardin des plantes in Paris. This bequest became part of the MNHN's collection when Alena died, joining Roger once more after six years of impatient solitude without him.

His Mineral Collection

Biographers and experts in the field reluctantly and cautiously note that Caillois' mineral collection is worthy of the great museums. Roger Caillois wanted there to be a museum in Vichy of his collection of documents and stones, and an exhibition held at the Valery-Larbaud cultural centre from May 21 to June 25, 1975 was intended to lay the foundation for it. Local politics quickly put an end to this project however – the town, though known for its water, was not as minerally inclined as Roger would have liked.
Certain scholars believe that Caillois' collection had as many as 2000 items, whereas Odile Felgine, in her book on Caillois, writes about some 220 specimens (217 to be exact) of which 165 (167) were given to the state in 1988. While the majority of the minerals were then duly catalogued, a third of them were dispersed among the general collection, including major pieces like the one mentioned in the donation as Caillois' "first labradorite," a sample that truly initiated his mineral prose, much more so than his wonderment at a quartz which haunted him long after he had glimpsed it in Brazil during the Second World War.
The labradorite in question was found in 2012 and then definitively identified in 2019 by comparing it with the illustration published in *Natura Pictrix* in 1959. Years of meticulous searching in many public and personal archives also enabled us to find most of the other specimens which had been scattered among numerous drawers, such as the *Agate blessée*, which may be Caillois' most beautiful agate. It was found in an unmarked box after ten years of research which began the moment I was convinced that it had been included in the 1988 donation.
To date, half a dozen specimens from that 1988 donation have still not been found, including a green jasper and a native copper, each one weighing not less than 30 kg according to the notarised document. Another major specimen from this donation, a native silver from Norway which Caillois magnificently describes in *Pierres*, was catalogued in the national collection but exchanged not long after – without the required paperwork for deaccessioning – for a specimen of rhodochrosite from Colorado of questionable quality. Finally, the two manuscripts accompanying the donation, one of which, titled *Pierres anagogiques*, was Caillois' last and remained incomplete, were never catalogued and have indeed never been found.
This count, however, omits many specimens that remained with the heirs of Jan Byk, Alena's son from a previous marriage. The specimens were moved and were not stored properly which led to damage. Nicolas Bos, President, CEO, and Artistic Director of Van Cleef & Arpels gifted the ones he had acquired in 2017 to the Muséum national d'histoire naturelle through the École des arts joailliers and its president Marie Vallanet-Delhom. This donation included 1043 items, about 1000 minerals and 40 miscellaneous objects: medals, artworks, every day objects, and manuscripts including Caillois' own inventory of the collection. This inventory concluded in November 1978 with item 2216, an agate described as *aux papiers depecés* (with torn paper), added the day before the author's death. But at a certain point, Caillois had shifted his inventory by a factor of one thousand: item 2216 therefore corresponded to the 1216th and final item.

The total number of one thousand two hundred and twenty remains approximate, since certain items had been replaced by Caillois. Others were not catalogued, with each drawer containing superb forgotten or sometimes hidden items that were not included in the inventory due to his emotional attachment to them.

Two Inheritances and Their Recovery

The complete and accurate consolidation of the Caillois collection was begun as soon as I arrived at the Musée national d'histoire naturelle in 2006, starting with the 1988 donation. After much work, about fifty specimens were painstakingly recovered but a half dozen or so items, including two manuscripts, have still not been found. From 2017 onwards, I was working on the donation of the collection still in the hands of the heirs, which Henri-Jean Schubnel had told me about in 2012. Those works were officially catalogued in 2021. A lengthy conservation project under the direction of Erik Gonthier at the Musée de l'homme was followed by my equally paintstaking work of identifying the items and cataloguing them according to French museum standards in order to not repeat the approximations that occurred when the 1988 donation was inventoried. The consolidation of these two collections gives us a fairly coherent view of the former Caillois collection, which has now been largely reconstituted.
Based on my archival research, however, at least one hundred and forty identified specimens, including about forty major pieces, are missing, having been scattered here and there after Caillois' death. In addition, the catalogue of this collection, while invaluable, is often too vague to identify many items whose inventory numbers created by Caillois were unfortunatly either not recorded at the time of the donation, or not marked on the samples themselves. Moreover, Caillois only noted the name of the dealer and the price of his acquisitions, but omitted the specimens' provenance or geographical origin. These omissions, which are hardly mineralogical, do not help in finding speciments within the inventory.
The Caillois collection has recently benefited from a great deal of scientific research by mineralogists, curators, and historians, which it had lacked in the not-too-distant past. As the study of the collection progressed, from biographical research to the study of specimens and written texts, the names attached to certain stones posed increasing problems: many of them appear neither in any of his publications nor in his catalogue, or when they do, the names are different. For example, an aggregate of calcite crystals from the 1988 donation is still called *Femme à l'enfant* (Woman with Child) at the Museum, and it is listed as coming from India. But I found it explicitly described in a chapter of *Pierres réfléchies* (1975) with the much more evocative name of *Officiant*. The *Woman with Child* was, in fact, a man in a traditional Japanese theatre costume! Caillois' inventory also allows us to confirm that it is a calcite from Mexico, and not India. This is just one example among many. I have concluded from this that it was common practice at the Museum to invent apocryphal names for certain specimens. It was therefore necessary to elimnate these posthumous misrepresentations, or "pseudo-Caillois'" as I like to call them, the equivalent of the *anhang* to German musicologists.
When Caillois gave names to his stones, he could be misleadingly figurative, even enigmatic. He quickly departed from his early idea of "pierres de rêve" or "dream stones" in which many still see landscapes, objects, or creatures. The names coined by Caillois were usually subtle, as they were linked to his extensive literary erudition. A study of the manuscripts reveals, however, that naming tends to disappear in Caillois' work from the 1970s onwards. For example, he named roughly half of his figurative specimens in *Écriture des pierres*, as can be seen by studying the captions for the illustrations. While within the exhibition presented by Caillois in Vichy in 1975, only six of the

twenty or so minerals on display were named. The professorial discourse and "majestic" iconography of the 1960s increasingly gave way to more effusive prose.

A Retrospective

The Caillois mineral collection remains one of modest size, put together by a talented amateur. It is neither a classical mineral collection, with a diversity representative of the most emblematic types of minerals, nor an expert grouping made up of rare specimens or put togther according to a distinctive mineralogical theme focused on a single place of origin or a particular type. We do not find in it rare specimens, "Rosetta stones" to support some new theory, gems that would impress the world's finest jewelers, nor "aesthetic-commercial" specimens aimed at investors or important mineral collectors such as Roger Titeux in France in his day. Caillois' collection must have been of fairly limited interest to Titeux, as Caillois, as a writer, was clearly neither materially nor naturalistically in competition with great collectors, despite the considerable financial resources at his disposal. Such collectors and Caillois even ignored one another, with mutual disdain.
The interest in the Caillois collection lies elsewhere. And it is enormous. At the heart of this interest, naturally, are the words, signs, dreams, and the prose that emerge from these crystals. So much attention had been dedicated to stones and so much had been written about them since Antiquity that taking up the challenge of reinventing the semantics associated with them was as daunting as it was audacious. Caillois triumphed nonetheless, lending them a noble pedigree that was both new and contemporary and that any great collector would be at pains to match, given how little attention they generally paid to the immaterial aspect of stones. What transitory messages could these minerals, rocks, and fossils have communicated to this Immortal who dreamed of eternity? These stones contribute to a universe all their own, a unique space so distinctive that it cannot be summed up in a way that captures its material, virtual, scientific, or literary magnitude. Like two-sided mirrors, at once synchronous and dissonant, they confound the literary and philosophical minds who have so meticulously – and patiently! – analysed the unique thought of Caillois.
But the vast majority of texts on the work of Caillois are quite evasive about or entirely ignore the poetically intriguing objects that the minerals, as seen through his unique lens as a writer, inspired. Some essays, while masterpieces of erudition, manage to distort the eternal illusion of minerals as inert with ever-increasing vital movement: from Caillois' text *Soleils Inscrits* (1965, p. 15, reprinted in *Pierres* in 1966), some of them have retained the description of a tree, with its circulating sap, from its budding shoots to its lifeless twigs. But these misleadingly biological words are just an incidental mineralogical projection, since, in the imagined reality, the foliage is a planturous quartz crystal with suggestive edges and transparencies that few people have been able to appreciate, much less actually see in real life, sparkling with crystalline majesty. Why do these scholars devote such scant attention to Caillois' stones? Why such disdain for the incredible light which bathes these minerals, inspiring the writer's poetic imagination?
And yet, from sociological paradigms to dream-like prose, Caillois lends inestimable value to his minerals, which he bequeaths to us in turn. Caillois, in his search, follows in the footsteps of the likes of Leondaro da Vinci, George Sand, and Pablo Neruda before him, who enchanted and were enchanted by minerals. Not to mention André Breton, ever-present and aways ready to pounce, Hans Arp, and many others, or those who, less skilfully, merely spoke of sad stones or praised gleaming jewels, from Eugène Sue to Madame de Genlis. Caillois knew better than anyone how to express his feelings, the complex reflections of a sociologist who dissected his subject at a rarely attained level of

thought, constantly verging on semantic excess as he poured over samples that, to novices, collectors, or even scientists, could at times seem austere.
The rigour with which Caillois selected his specimens is expressed in distilled prose. It is here that Caillois the scientist and sociologist, rather than the mineralogist, expresses himself most complexly. It is also here that Caillois the poet expresses himself most intensely. Far from being dry and formulaic (I dare not say lapidary), Roger Caillois unlocks a strange passion, one that is entirely his own but that still influences many of today's mineralogists, unbeknownst, perhaps, to them. Stones suggested astonishment to him, neurosis even: Caillois is innovative in the way in which he dissects the mysteries of his fascination. From *Minotaure* to *Inquisitions*, Caillois learned from looking at minerals as he distanced himself from André Breton – whom he was not in close contact with but never entirely lost sight of – to join other milieus. And yet, Caillois remains indebted to Breton for, among other things, his *maison que j'habite, ma vie, ce que j'écris* ("the house I live in, my life, what I write" from *La beauté sera convulsive*, 1934), that is, for a contemporary sanctification of the mineral realm as a source of artistic if not poetic inspiration. For the French, trained from adolescence onwards to have an atavistic repulsion for stones ("une répulsion atavique du caillou"), this was no mean feat: the Chinese had already incorporated it under the Song Dynasty in the tenth century and the Italians had done so since the Renaissance, if not from Antiquity, as we may note from the agate from Pyrrhus that Caillois describes, copying again and again from the text of Pliny the Elder dating to the first century A.D. But Caillois tried to carefully situate himself between *cailloux de poètes* and *échantillons de minéralogistes*, between the stones of poets and mineralogists' samples. He delved deeply into scientific fact, venturing quite freely into that domain, well aware of the risks involved. Demonstrating his familiarity with the most abstuse of words, Caillois liked to play at being a mineralogist: he liked to call himsef a *mineralogue*, a word that had hardly been used since the eighteenth century. In reality, he was more of a mineralophile than a mineralogist. Caillois wrote about, lectured on, and proclaimed and expresssed his love: he learnedly transmitted the principles of what motivated him in his heavily redacted manuscripts and, above all, in his illustrated talks. From enhydro agates, to flourines, to the age of gogottes and the research that went into his text *Château*, Caillois could sometimes implicate himself, carried away by spontaneous discourse that was lethal from a scientific point of view. His solid dreams, like waking dreams that have become real, still manage to infuse a kind of *estro* which in turn becomes, in the poetry of stones, a universal "lesson in things." Beyond the phantasmagorical digressions with which he sought to dazzle impervious matter, Caillois' minerals have become sumptuous guides, which will exhilarate us if we pay attention. For it is the vibrancy of the signs that matter here and that will carry us through.

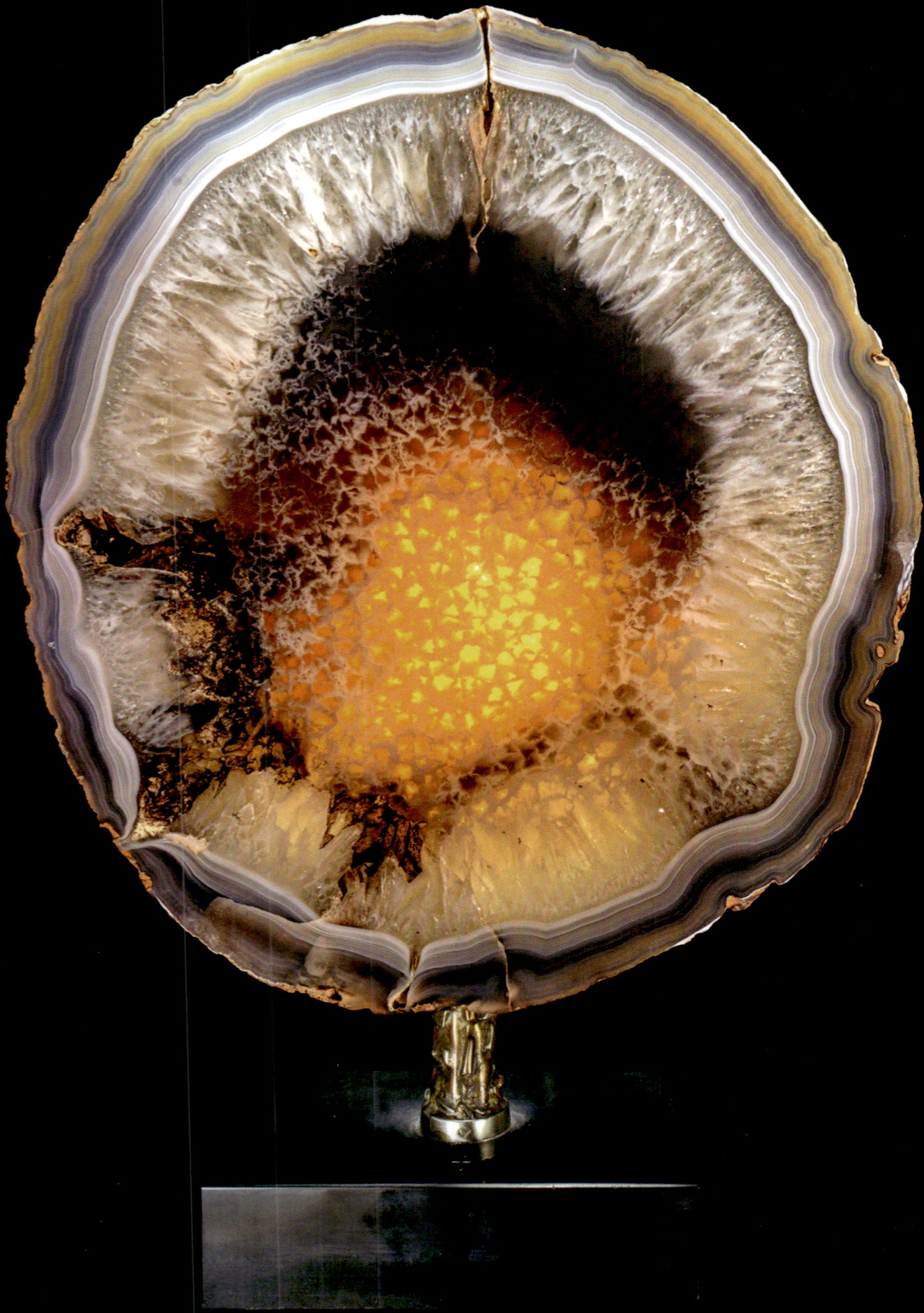

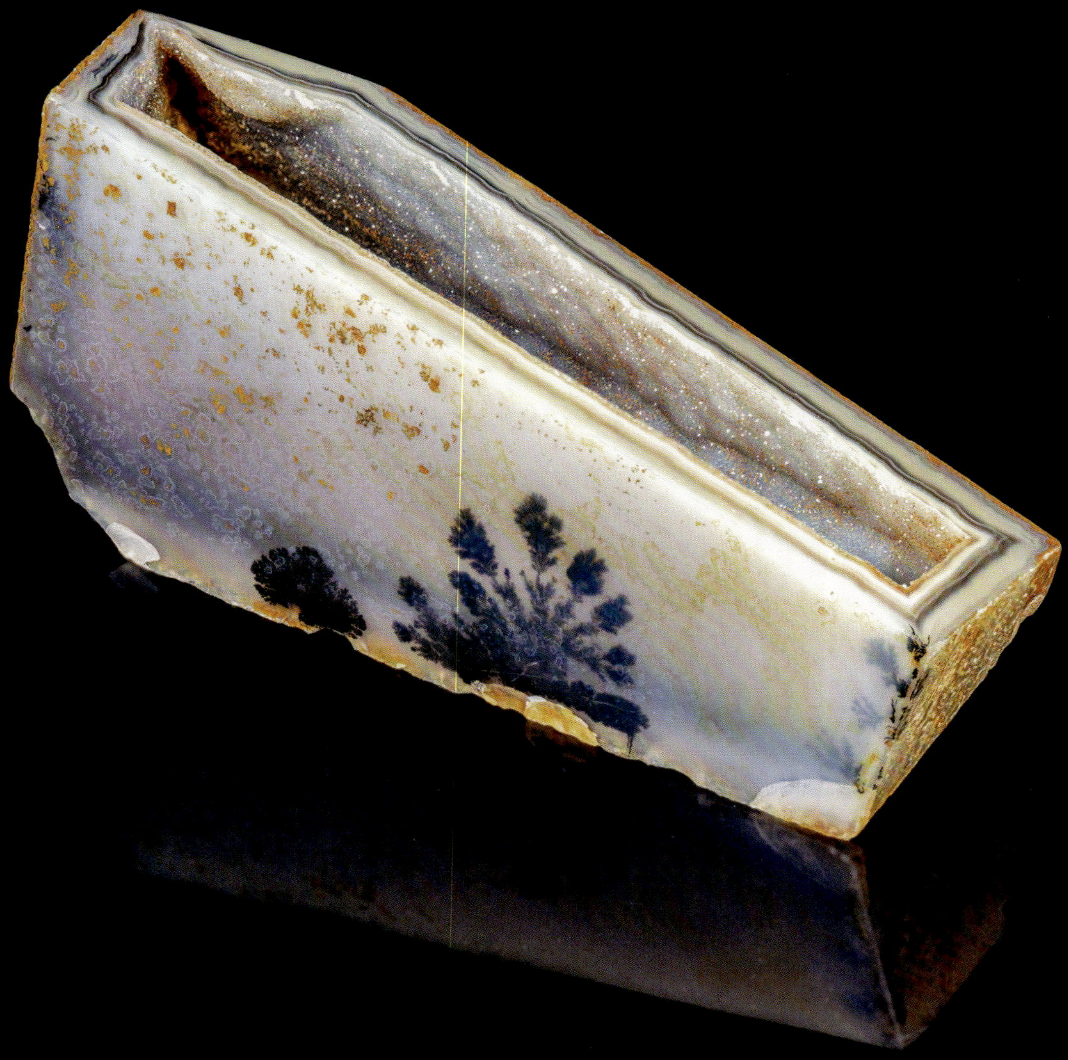

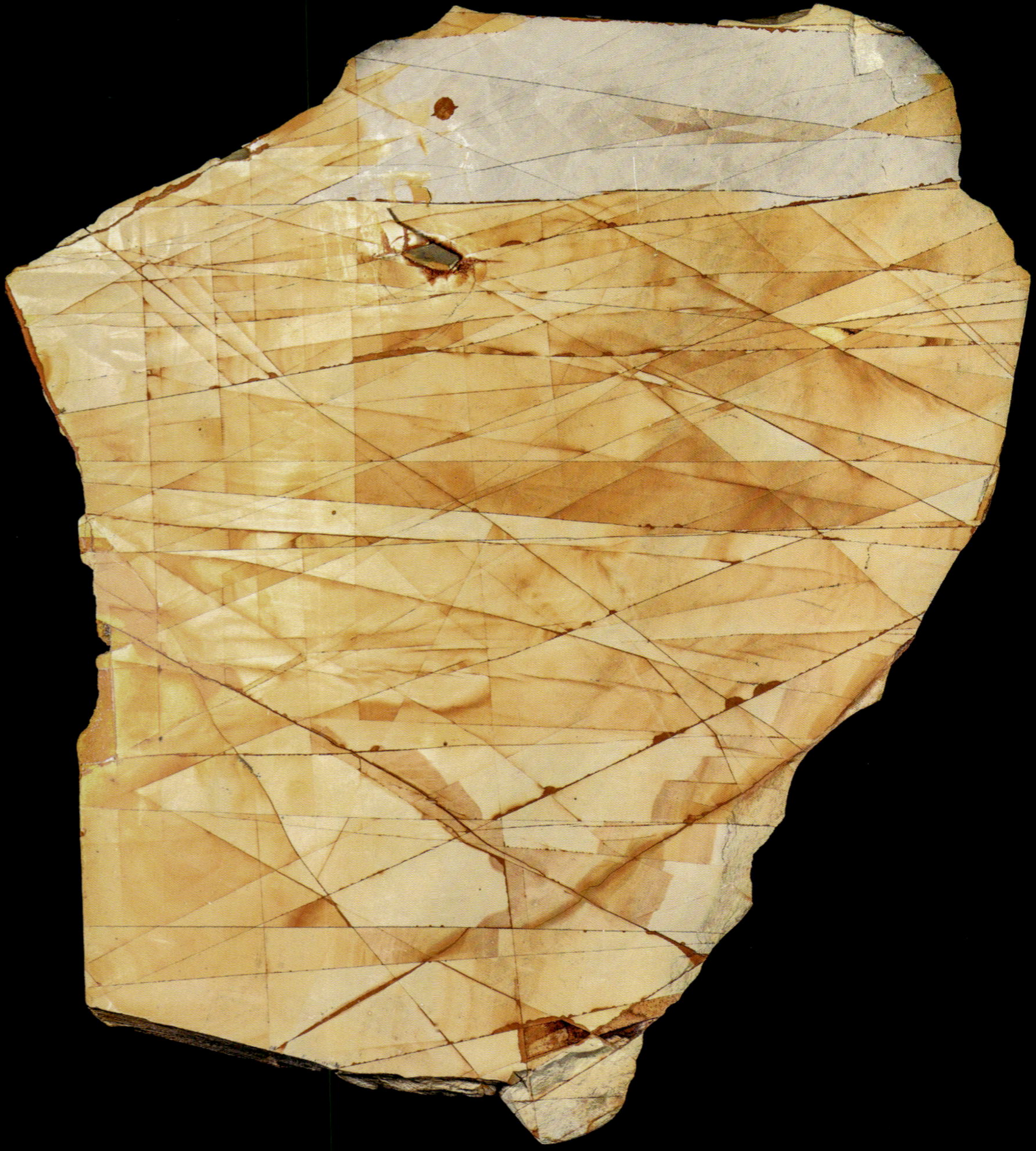

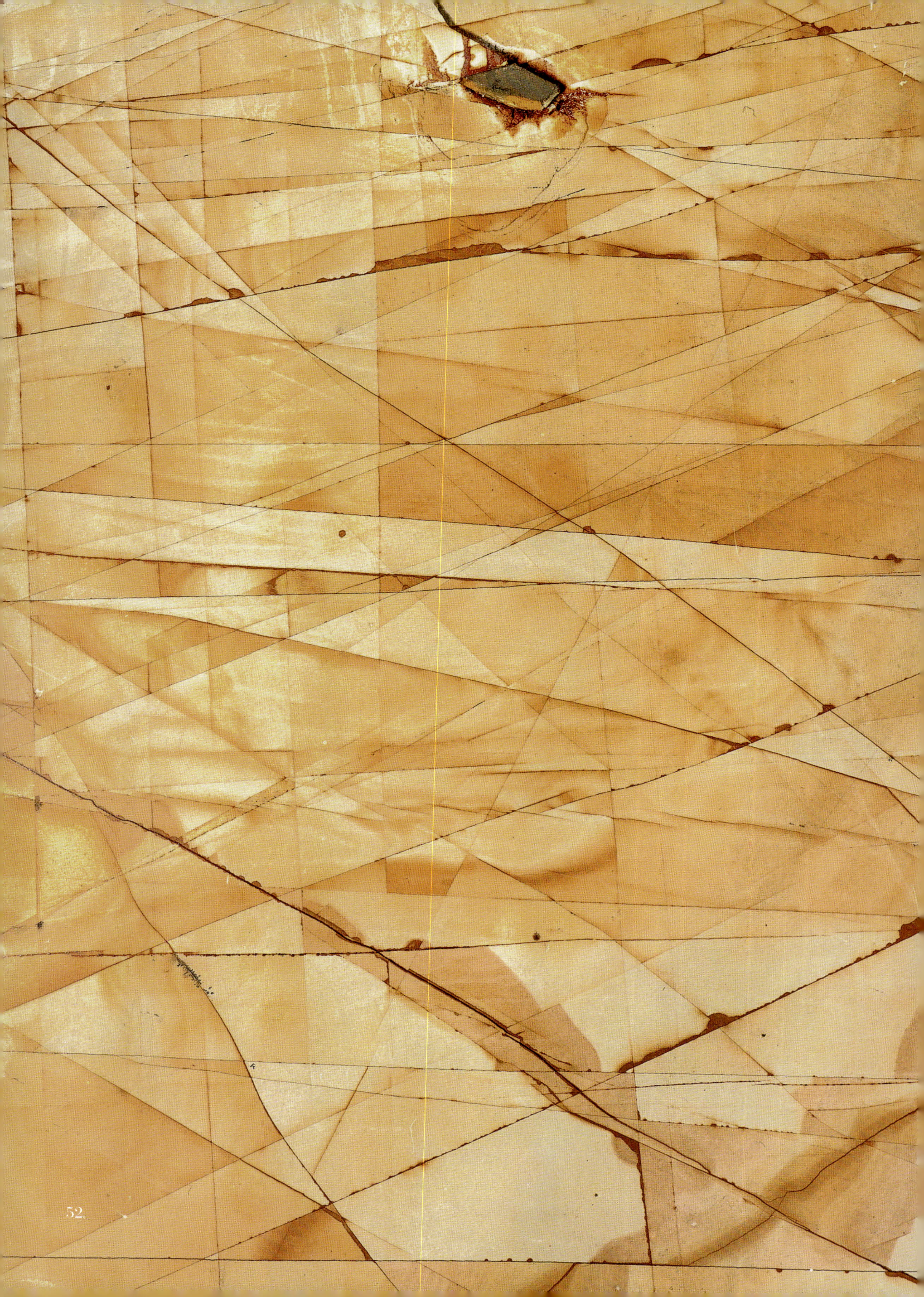

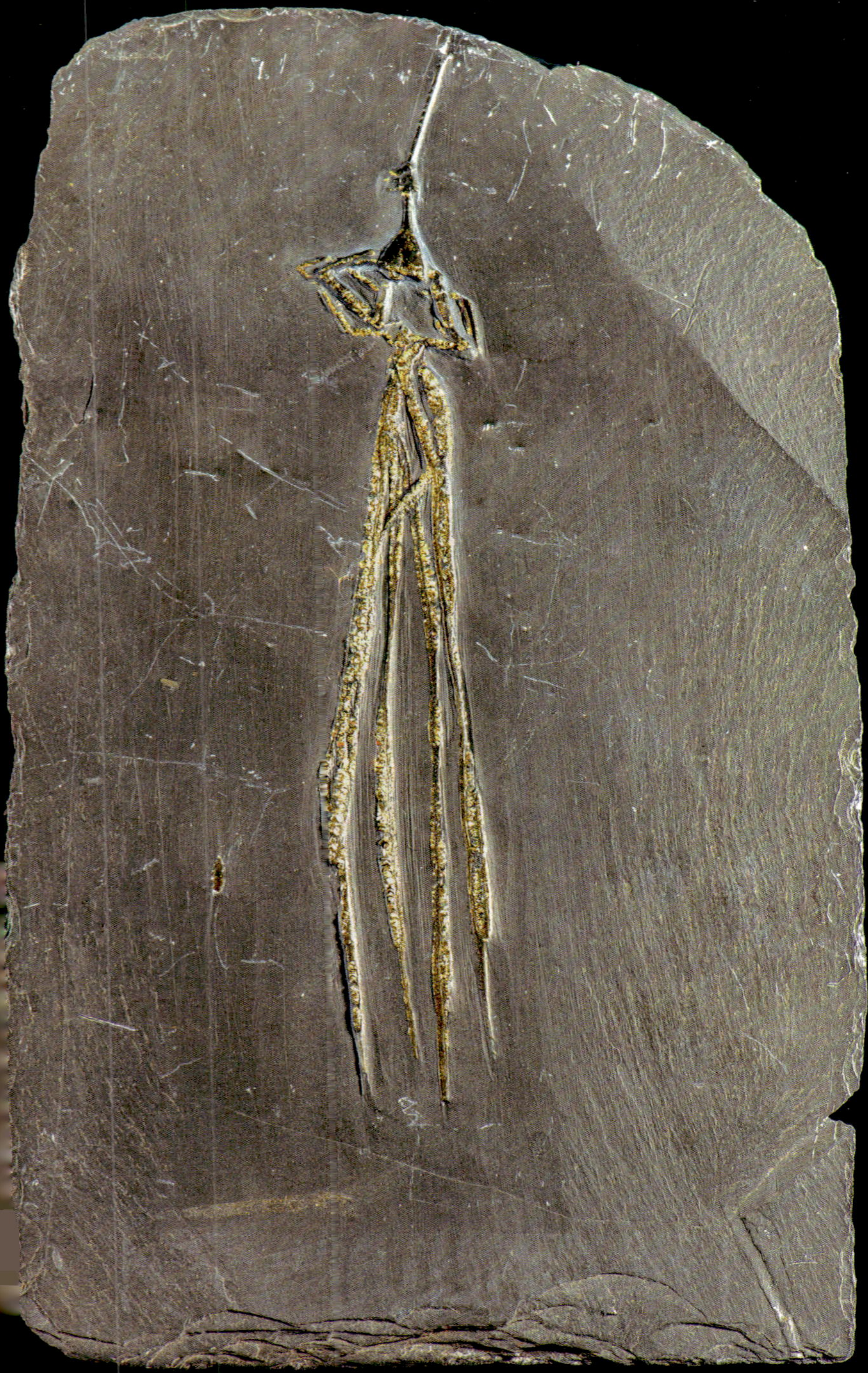

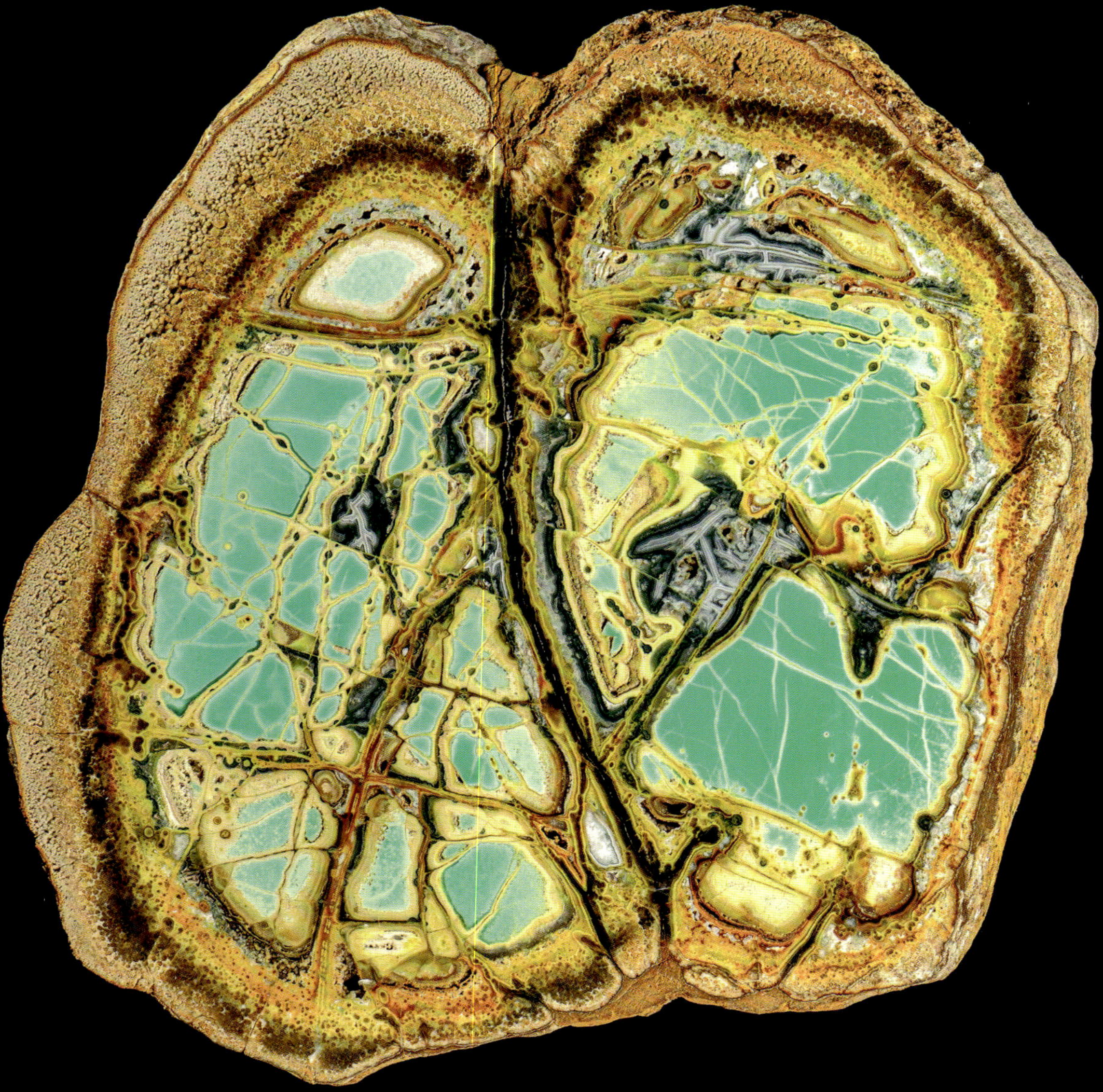

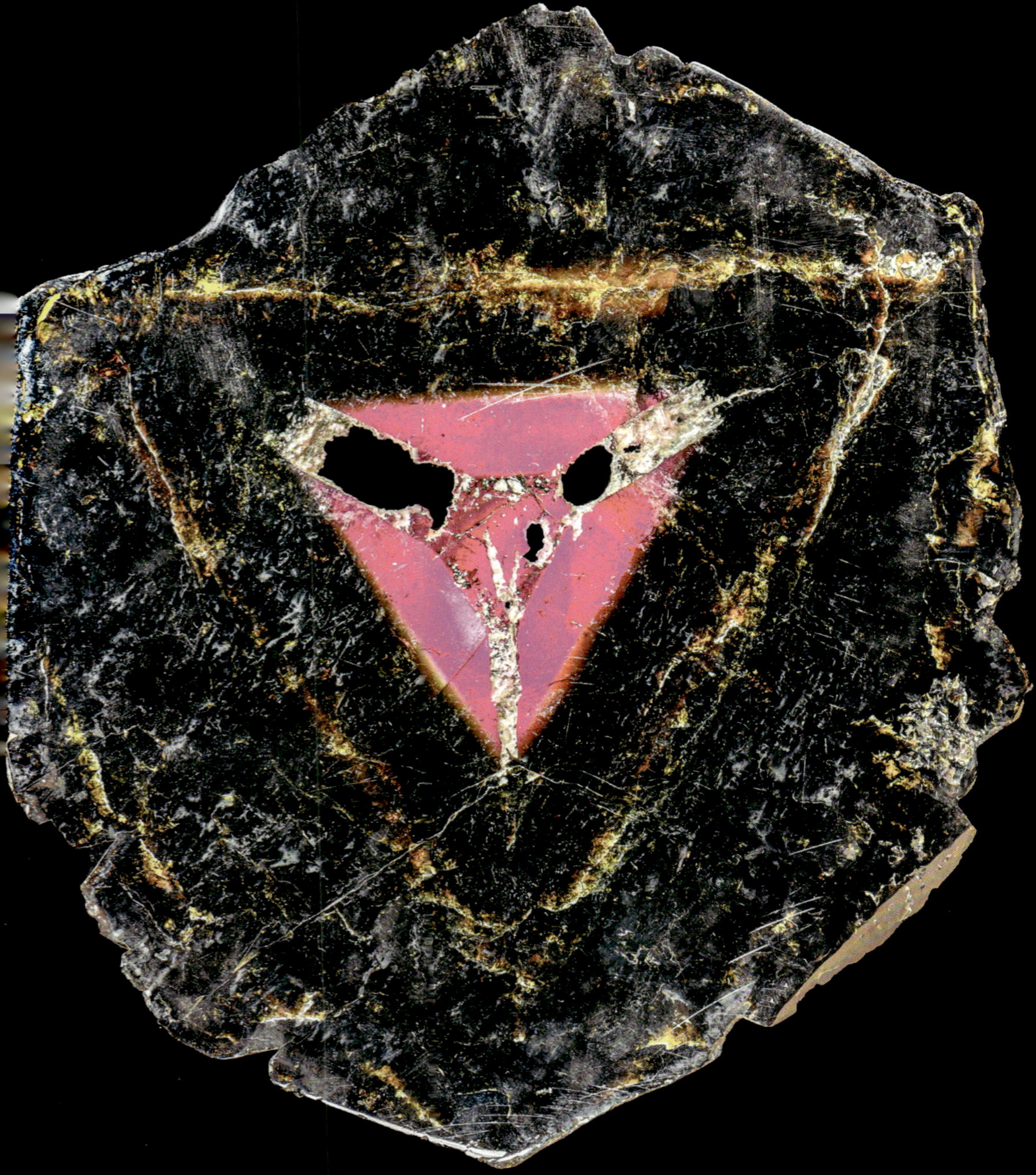

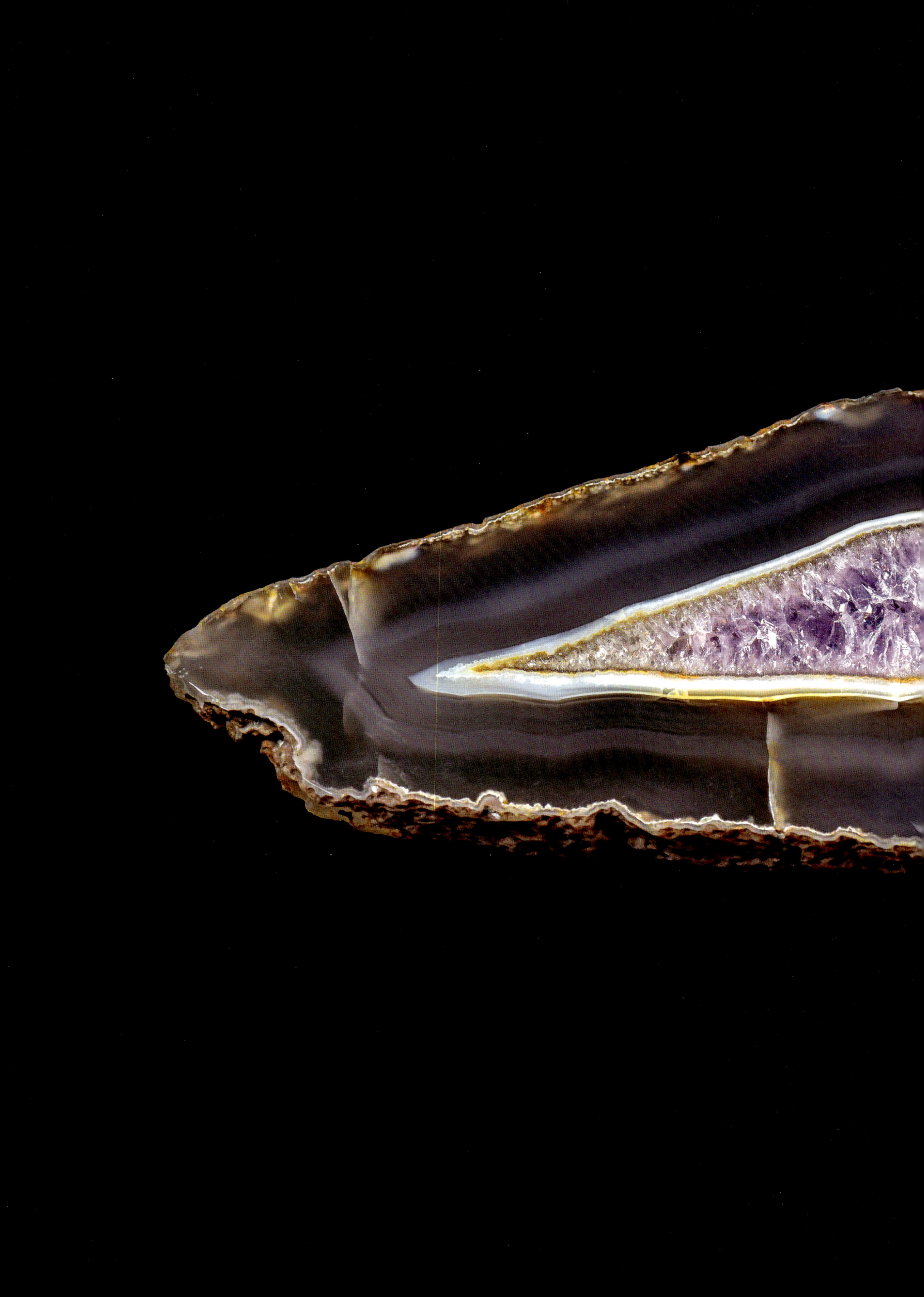

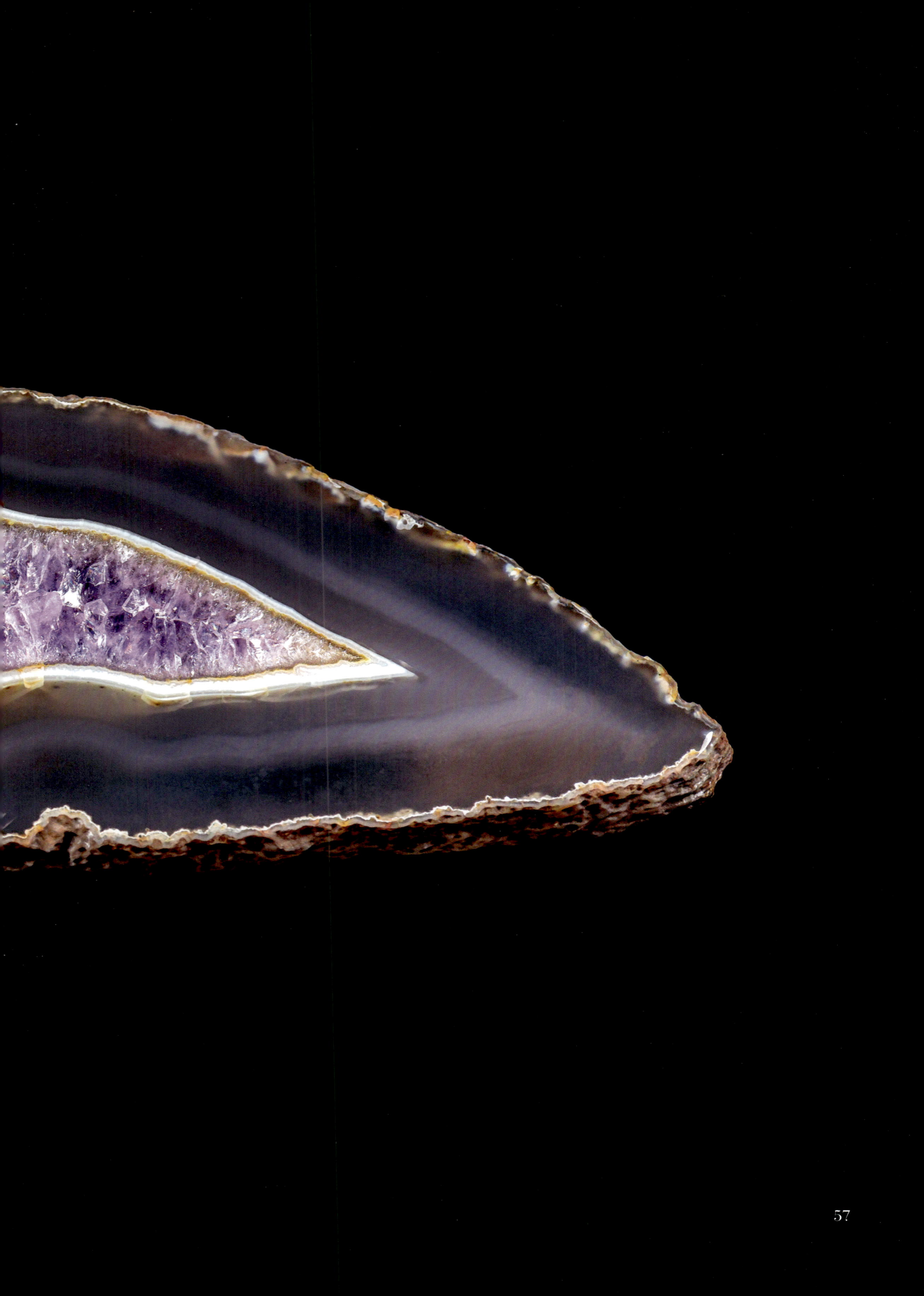

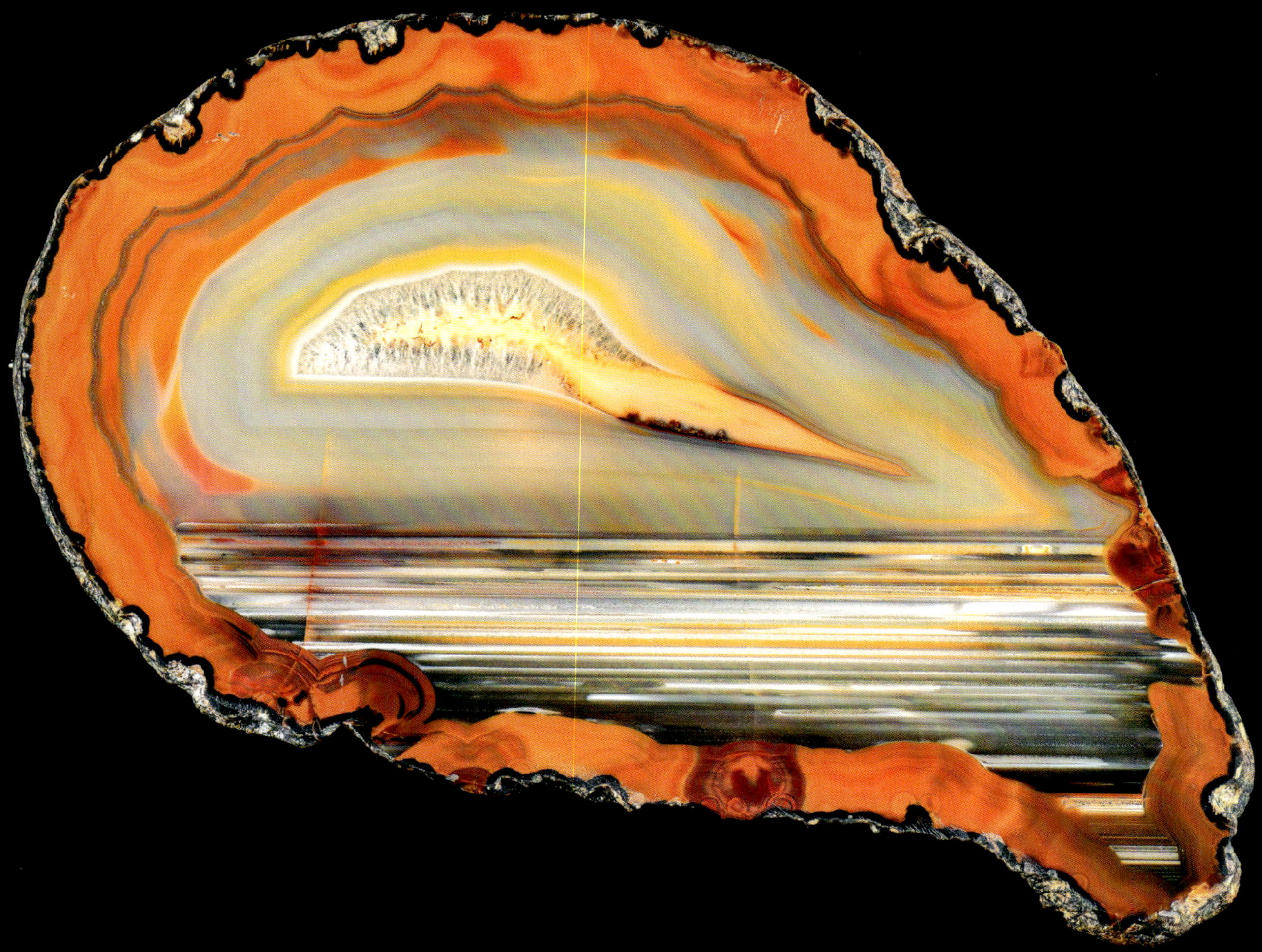

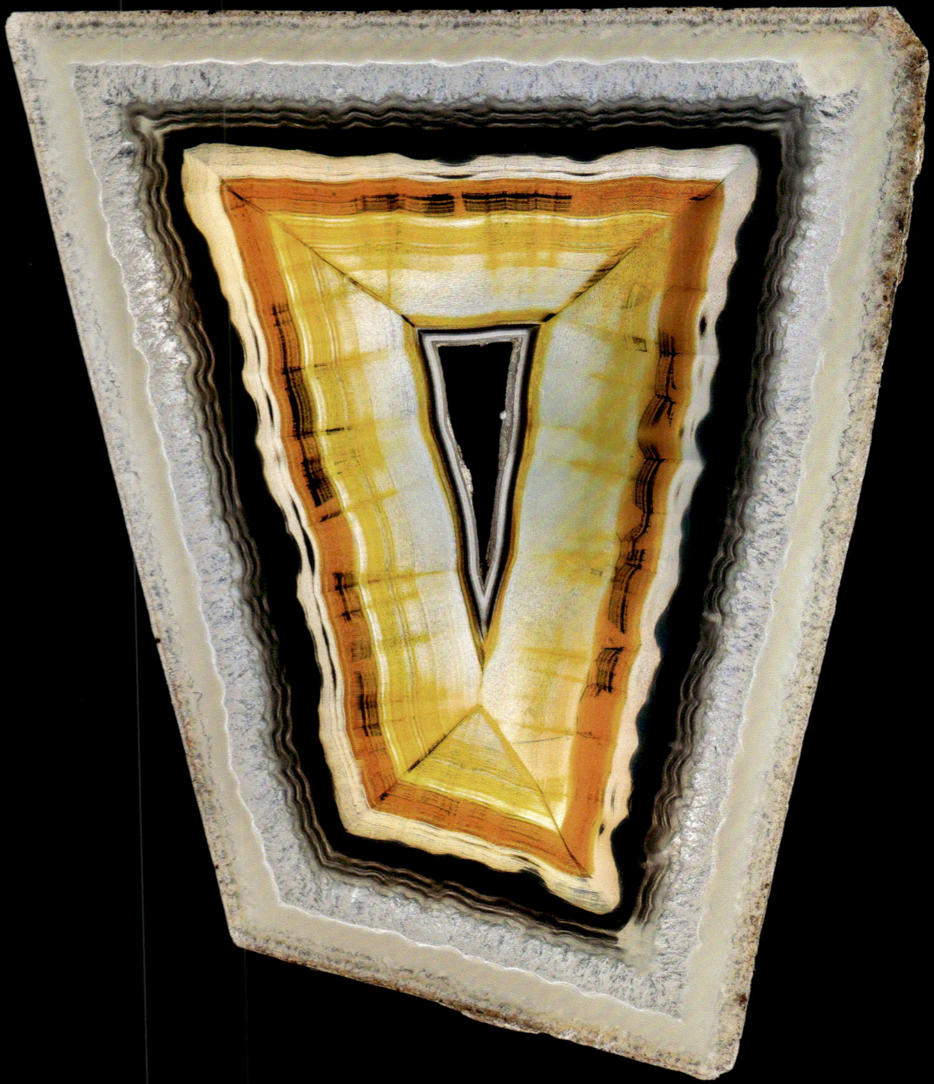

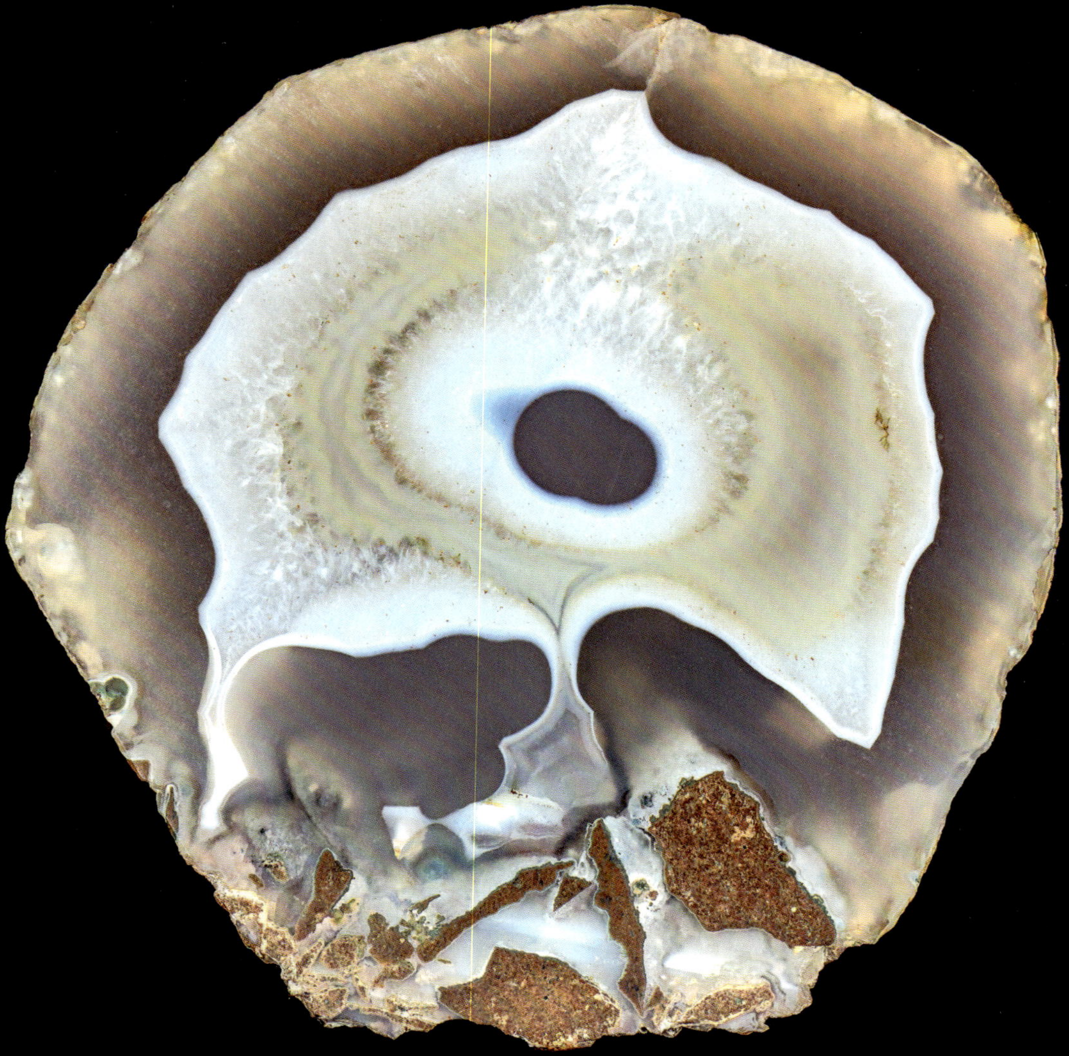

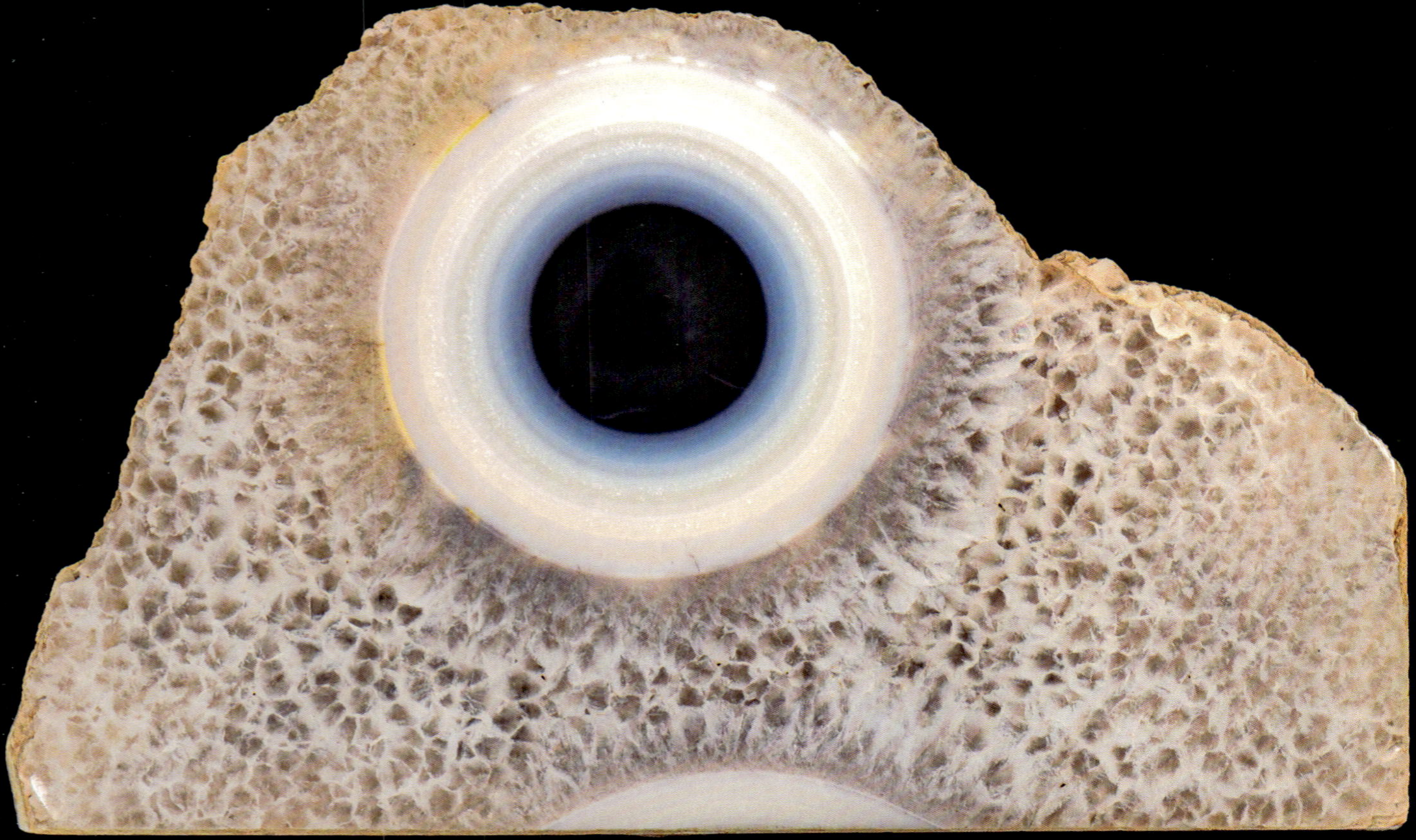

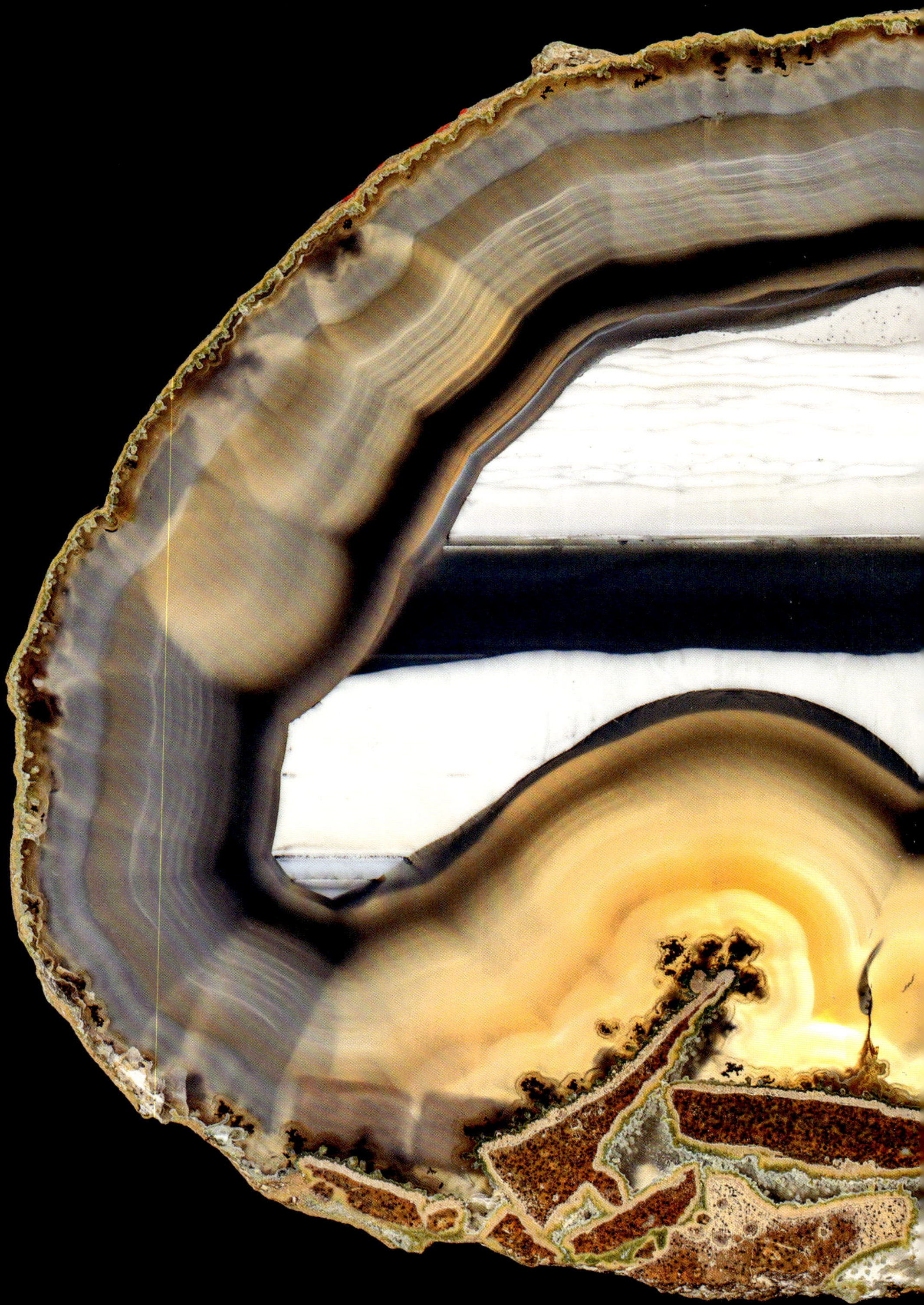

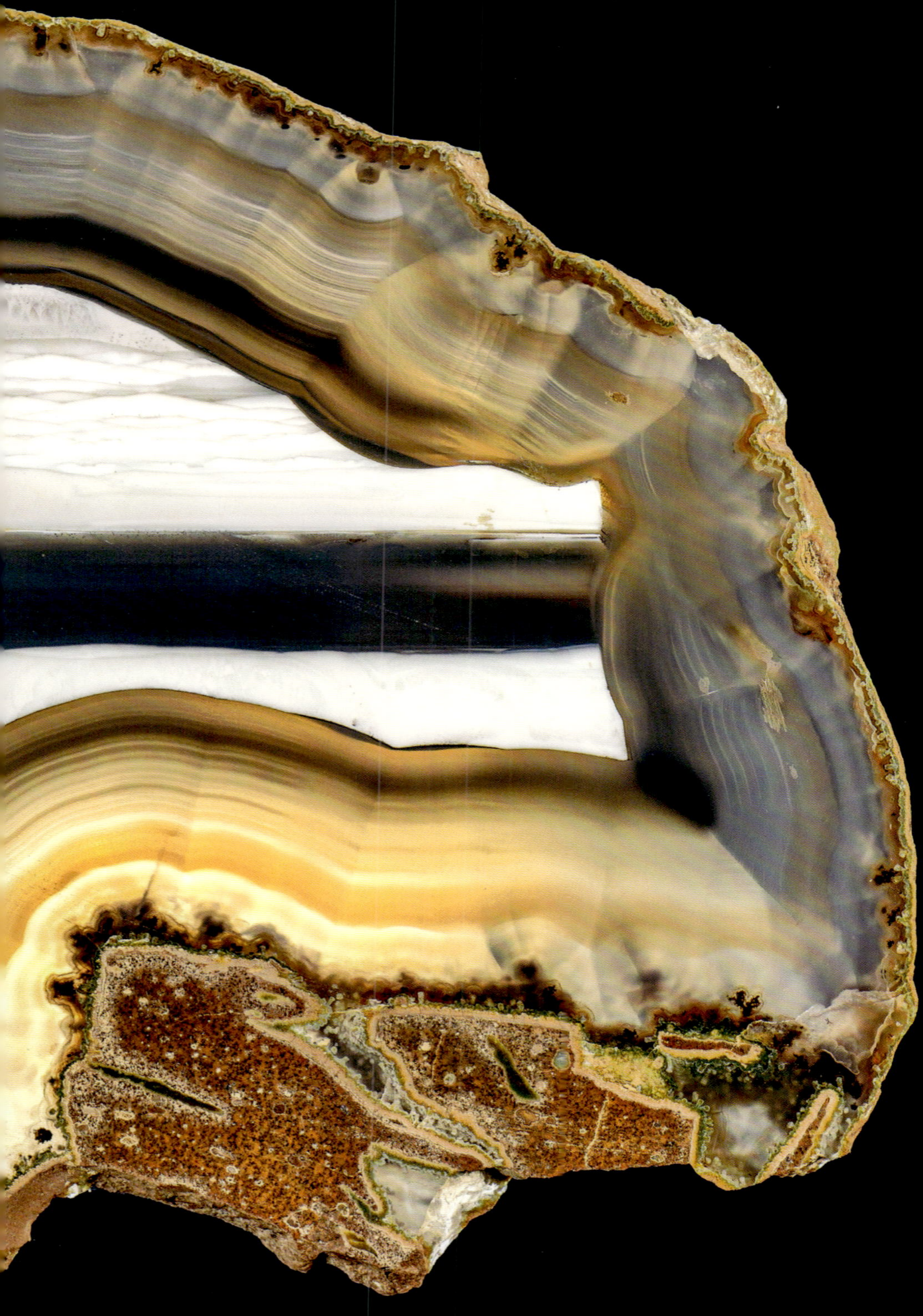

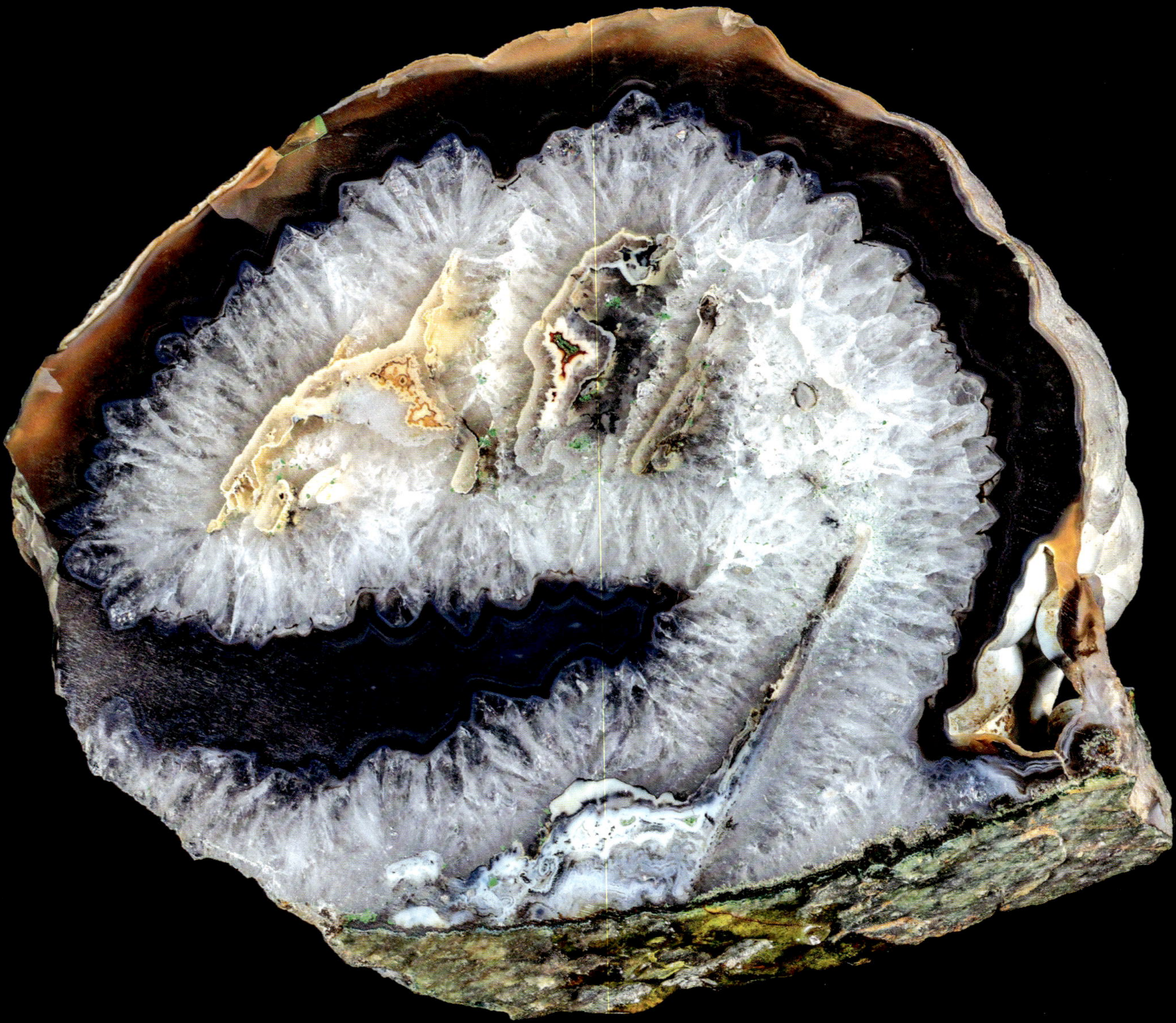

The Writing of Stones

by Roger Caillois

Mechanically testing stones matters less than clearly perceiving their qualities: their very appearance speaks to us on a basic level, as does the way they let us read them, regardless of their specific physical features. Franco Maria Ricci, who was a geologist, may perhaps have guessed that this kind of bond was a crucial link in the chain between the mineral kingdom and the domain of art. Roger Caillois — an "Immortal," that is to say, a member of the Académie Française — who collected and could read *stones, thought so too. The use of the verb "read" may seem polysemic: we can read a book, a tarot deck, the future, the human soul... Roger Caillois could read stones in all these ways, at the very least, and most likely in others still.*

The Image in the Stone

Just as men have always sought after precious stones, so they have always prized curious ones, those that catch the attention through some anomaly of form, some suggestive oddity of colour or pattern. This fascination almost always derives from a surprising resemblance that is at once improbable and natural. Stones possess a kind of *gravitas*, something ultimate and unchanging, something that will never perish or else has already done so. They attract through an intrinsic, infallible, immediate beauty, answerable to no one, necessarily perfect yet excluding the idea of perfection in order to exclude approximation, error, and excess. This spontaneous beauty thus precedes and goes beyond the actual notion of beauty, of which it is at once the promise and the foundation.
For a stone represents an obvious achievement, yet one arrived at without invention, skill, industry, or anything else that would make it a work in the human sense of the word, much less a work of art. The work comes later, as does art; but the far-off roots and hidden models of both lie in the obscure yet irresistible suggestions in nature. These consist of subtle and ambiguous signals reminding us, through all sorts of filters and obstacles, that there must be a preexisting general beauty vaster than that: perceived by human intuition – a beauty in which man delights and which in his turn he is proud to create. Stones – and not only they but also roots, shells, wings, and every other cipher and construction in nature – help to give us an idea of the proportions and laws of that general beauty about which we can only conjecture and in comparison with which human beauty must be merely one recipe among others, just as Euclid's theorems are but one set out of the many possible in a total geometry.
In stones the beauty common to all the kingdoms of nature seems vague, even diffuse, to man, a being himself lacking in density, the last comer into the world, intelligent, active, ambitious, driven by an enormous presumption. He does not suspect that his most subtle researches are but an exemplification within a given field of criteria that are ineluctable, though capable of endless variation. Nonetheless, even though he neglects, scorns, or ignores the general or fundamental beauty which has emanated since the very beginning from the architecture of the universe and from which all other beauties derive, he still cannot help being affected by something basic and indestructible in the mineral kingdom: something we might describe as lapidary that fills him with wonder and desire.
This almost menacing perfection – for it rests on the absence of life, the visible stillness of death –appears in stones so variously that one might list all the endeavours and styles of human art and not find one without its parallel in mineral nature. There is nothing surprising about this: the crude attempts of that lost creature, man, could not cover more than a tiny part of the aesthetics of the universe. No matter what image an artist invents, no matter how distorted, arbitrary, absurd, simple, elabourate, or tortured he has made it or how far in appearance from anything known or probable, who can be sure that somewhere in the world's vast store

there is not that image's likeness, its kin or partial parallel? Even setting such similarities aside, human beings are attracted and amazed by many mineral formations: spiny tufts of quartz; the dark caves of amethyst geodes; shiny slabs of variscite or rhodochrosite agate; fluorine crystals; the golden, manysided masses of pyrites; the simple, almost unsolicited curve of jasper, malachite, or lapis lazuli; any stone brightly coloured or pleasingly marked. Connoisseurs, in such cases, admire the qualities of a material that is constant and unchanging: purity, brilliance, colour, structural rigour – properties inherent in each kind and present in every example. Their values are intrinsic, without external reference. The price a purchaser pays for them depends on weight, rarity, the amount of work involved, just as with a length of satin or brocade, a bar of refined metal, or a gem. Like such commodities, these stones are exchangeable, since there is no difference between one of them and another example of the same kind, size, and quality.

The whole picture changes when singularity is what is sought after. The stone's inherent qualities and special geometry are no longer of primary concern, perfection no more the sole or even the main criterion. This new beauty depends much more on curious alterations brought about in the stone itself by means of metallic or other deposits, or on changes in its shape due to erosion or serendipitous breakage. Some pattern or peculiar configuration appears in which the imaginative observer descries an unexpected, in this context an astonishing and almost shocking copy of, an alien reality. Such semblances emerge from their long concealment when certain stones are split open and polished, presenting the willing mind with immortal small-scale models of living beings and inanimate things. Admittedly such marvels are the result of mere chance, such resemblances only approximate and dubious, occasionally farfetched or even arbitrary. But once perceived they soon become tyrannical and deliver more than they promised. The observer is always finding fresh details to round out the supposed analogy. Such images *miniaturize*, for his benefit alone, every object in the world, providing him with stable duplicates which he may hold in the palm of his hand, carry about from place to place, or put in a glass case. Moreover, such a duplicate is not a copy; it is not born of an artist's talent or a forger's skill. It has been there always: we only had to find our way into its presence. Ordinary rocks as well as various types of mineral specimens make up this prey of Pan. In China, poets and painters would see in a cleft stone a mountain with its peaks and waterfalls, its caves and paths and chasms. Collectors ruined themselves to possess crystals in whose translucent depths they discerned mosses, grasses, and boughs laden with flowers or fruit. An agate may shadow forth a tree, several trees, groves, a forest, a whole landscape. A piece of marble can suggest a river flowing among hills; the clouds and lightning flashes of a storm, thunderbolts and the grandiose plumes of frost; a hero fighting a dragon; or a great sea full of fleeing galleys, like the scene the Roman saw reflected in the eyes of the Eastern queen already planning to betray him.

One kind frequently depicts a burning town, with its towers and steeples and campaniles crashing down. On the agate of Pyrrhus antiquity made out Apollo with his lyre, surrounded by the muses, each with her special attributes. In the seventeenth century Gaffarel, Richelieu's librarian and the king's chaplain, devoted a whole weighty volume to *gamahés*, healing talismans made of stones inscribed with natural astrological hieroglyphs.[1] Princes and bankers of the same period collected unusual specimens sought out for them at great expense by the numerous agents of specialist merchants. Learned men, among them Aldrovandi and Kircher, divided up these marvels into families and types according to the images they managed to distinguish in them: Moors, bishops, lobsters, streams, faces, plants, dogs, fishes, tortoises, dragons, death's heads, crucifixes – everything a mind bent on identification could fancy. The fact is that there is no creature or thing, no monster or monument, no happening or sight in nature, history, fable, or dream whose image the predisposed eye cannot read in the markings, patterns, and outlines found in stones. The more unusual, definite, and undeniable the image, the more the stone is prized. Stones that offer rare and remarkable likenesses are regarded as wonders, almost miracles. They should not exist, and yet they do, at once impossible and inescapable. At the same time they are treasures, the result of thousands upon thousands of chances, the winning number in an infinite lottery. They owe nothing to patience, industry, or merit. They have no market rate or price. Their value is not commercial and cannot be calculated in any currency; it laughs both at the gold standard and at purchasing power. It is not convertible into labour or goods. It depends solely on the covetousness, pride, and competitiveness generated by the desire to possess or the pleasure of possessing them. Each stone, as unique and irreplaceable as a work of genius, is a valuable at once pointless and priceless, with which the laws of economics have nothing to do.

That being the case, they are frequently regarded as amulets and talismans. The owner of such a wonder, brought into being, discovered, and delivered into his hands through an inconceivable concatenation of chances, easily comes to believe that it could have come to him only through some special intervention of fate. He grows passionately attached to it, thinks of it as a guarantee of health and success. But we are not concerned here with the magical virtues superstition attributes to stones inscribed with images, nor even with the joyful gloating of their owners.

In some Eastern traditions insight may be obtained from the strange shape or pattern in a gnarled root, a rock, a veined or perforated stone. Such objects may resemble a mountain, a chasm, a cave. They reduce space, they condense time. They are the object of prolonged reverie, meditation, and self-hypnosis, a path to ecstasy and a means of communication with the Real World. The sage contemplates them, ventures into them, and is lost. Legend has it that he never returns to the world of mankind: he has entered the realm of the Immortals, and become an Immortal himself.

I leave such fabulous emotions to their fate, concerning myself only with the present-day competition between the products of nature and the works of painters and sculptors, whether nature's offerings appear to represent something or are pure sign. Artists have added to them or merely signed and framed them as they were, thus putting them in the same category as pictures. So they must have found some correspondence or element in common between their own creations and these works executed by no one. The juxtaposition should be revealing. At the very least it brings out some strange reversals. We have seen modern painters first give up trying to reproduce their models exactly, then abandon models altogether and eschew any kind of representation. And those markings on stones are considered most interesting which do not represent anything. Yet at the same time those rarer stones that do seem to depict something are coming back into favour. It is strange to think that nature, which can neither draw nor paint any likeness, sometimes creates the illusion of having done so, while art, which has always been successful at resemblances, renounces its traditional, almost inevitable and "natural" vocation and turns to the creation of such forms as nature itself abounds in mute, unpremeditated, and without a model. This inversion in the order of things seem simultaneously to reveal and to conceal a problem. To clarify the data of the problem, though I do not promise to find its solution, I shall try to define the ways in which nature sometimes gives us the impression of representing something. I should also like to explain the strange attraction of these manifestly illusory likenesses.

"Pierres-auxmasures" or Ruin and Landscape Marbles

Among the image-bearing stones which have long excited the envy of connoisseurs, likewise among the strangest specimens and thus the most sought-after for curio cabinets, two types of marble deserve special mention. They stand out first because they were frequently invoked as providing a crucial argument in favour of the theory of *lusus naturae*, or sports of nature, as against that of fossils, which explained the phenomenon involved as due to the petrification of ancient living organisms. Another reason for according them special treatment is that they were frequently polished and framed and treated like real pictures. The two types in question are the *pierres-aux-masures* (also known as ruin marbles or, internationally, as *paesine)* which come from a site near Florence, and the landscape marbles from a quarry near Catham, Gloucestershire in England, a site long ago filled in and covered with blocks of flats.

For scholars these marbles proved that nature, which through mere creative fantasy could depict ruined cities or pleasant valleys dotted with smiling groves, was even more capable of spontaneously producing images of fishes, mollusks, or ferns. They declined to take so-called imprints seriously, recalling the fact that such appearances of marine animals had often been found a long way away from the sea and sometimes even on mountain tops; moreover, they were different from known shells and fishes. This clearly meant they were whims of nature, just like the tiny ruined cities in the Tuscan marbles and the little rows of trees in those from England.

In Nuremberg in 1777 there was published, in a translation from the German, a work of great beauty and magnificence in several volumes somewhat ponderously entitled *Recueil des Monuments des catastrophes que le globe terrestre a essuyées* (Collected vestiges of the catastrophes suffered by the terrestrial globe.) It is a catalog of fossils and similar specimens and, to quote the rest of the title, "contains petrifications and other curious stones, drawn, engraved, and coloured after the originals, with descriptions."

The book, begun by Georg Wolfgang Knorr, was continued by Johann Ernst Immanuel Walch, counsellor at the court of the duke of Saxe-Weimar and professor of oratory and poetry at the University of Jena. This study, which firmly opposes the theory of *lusus naturae*, tracing its history and mocking its absurdities, is illustrated with plates the quality of which has never been equaled despite the technical progress made since. Some of the book's strangely shaped fish, printed in deep rich blacks, irresistibly bring to mind certain lithographs by Braque.

The book also contains a description, significant in this context, of ruin marbles:

a kind of stone found near Florence and named after the pictures found upon them. These represent ruined cities, towers and pyramids, crumbling walls and houses. The stone is commonly grey in colour, and the marks, which are slightly darker, imitate a birds-eye view. The ruins are brown, shaded darker brown in places. Between the ruins and especially near their base are little dendrites, as if the ruins were covered with moss. These stones vary in beauty and in degree of likeness to the subjects they represent. According to some authors such stones split very easily into layers; they are fine-grained and therefore take an excellent polish. The pictures go right through the thickness of each layer. Such stones must not be confused with the arborized marble *of Florence... The Italians call this latter* pietra emboscata, *and call the stone depicting ruins* pietra cittadina. *Lachmund's* Oryctographia Hildesheimensis *was the first to make this stone known in Germany, though by means of a very poor illustration. It was subsequently mentioned by Wormius, Kilentini, Besler, Kundmann, Davila, and several others. But most of them failed to give an exact description or inquire into the origins of the specimens. Brückmannhad something to say about them, and this should be collated with Jean Elie de la Faye, Keysler, and an anonymous article in the* Berlin Magazine.[2]

In 1733 Brückmann, who tried to explain the miracle and came down in favour of freaks of nature (rightly, in this case), appealed to some fluid mineral substance of various colours, converted into marble solidity through the action of a "coagulant and gorgonic" spirit.[3]

But these were not the absurd yet revealing ramblings of a science in search of its principles. In the late eighteenth century people were really only interested in explaining what was considered merely an odd phenomenon. The learned authors of the *Recueil* pointed out that cliff-strewn deserts, rocks, and debris of cities came into the category of "accidental figures, ciphers, and suchlike marks, by which formerly people set great score, but which have now lost all their credit because these Sports of Nature are regarded as no more than a curious amusement."[4]

The scholars had spoken. But the similarity between the Florence marbles and the works of painters still went on troubling men's minds, as is shown by the following extract from, paradoxically enough, a *Natural History of Minerals*, which was commonly appended to early nineteenth-century editions of Buffon and in which the art critic's attitude almost eclipses that of the mineralogist. Some of these stones, writes the author, *show, against a yellowish-white background, black dendrites in the shape of trees and shrubs arranged in several rows and in different planes, pleasantly imitating a landscape. This has caused this kind of stone to be called* alberese, *or* marmo paësino: *it is hard, compact, and takes a very fine polish.*

The other variety also has pictures of landscapes, but these represent ruins. They show ramparts, towers, steeples, obelisks, buildings of all kinds, but half in ruins and surrounded by debris. These ruins are dark in colour like a bistre drawing; the background or sky is lighter, being yellowish or bluish grey, against which the ruins stand out the more clearly because they are surmounted by a white band that makes them seem to be lit by the setting sun.

This white band sometimes forms long points tipped with a reddish hue, a very good likeness of the flames in a conflagration.

The sky has vaguely wavy, twisted lines which resemble clouds, and is often dotted with round black marks, like bombs launched against a besieged city. In the foreground of these pictures, especially when they are of some size, there is commonly what painters call a terrace, a strip of ground strewn with shrubs and undergrowth which completes the resemblance to a painted landscape.

All these charming accidents, in which nature so well imitates the productions of art, please by their singularity even those who are least interested in mineral productions. But they especially interest the naturalist, and above all arouse his curiosity to discover their causes.[5]

In fact, we know now, but only now, that the images on these stones could be more accurately described as a geological anticipation of the lofty machicolations of Manhattan, or, more precisely, of Bernard Buffet's stark New York panoramas with their dominant verticals.

There was a considerable trade in Florence marbles in the sixteenth and seventeenth centuries. Philipp Hainhofer, an Augsburg merchant, was involved in it and supplied Philip II, duke of Pomerania, in 1617, and Gustavus Adolphus, king of Sweden, in 1632, with cabinets decorated with that kind of stone. The cabinet belonging to Rudolph II, in Prague, was, according to his physician Anselm Boetius de Boor, decorated with admirable specimens, which looked "like paintings rather than stones." In his book *Aberrations*, Jurgis Baltrusaïtis gives many examples of public and private collections which contained such items, polished, mounted on slate, and set in often luxurious frames.[6] There. is no doubt that they were regarded as genuine works of natural art. Thus catalogs of the period record that the Settala Museum in Milan possessed two slabs taken from a stone sawn through the middle and showing the burning of Troy "in so natural a manner that Zeuxis could not have painted it better." The Cospi Gallery in Bologna had a marble depicting the same subject: "a white marble veined with red, dark, and other colours so disposed by Nature that they call to mind the woes of Troy, with the city and the rocks in flames."[7] Nowadays this type of "picture" has emigrated from the art gallery to the museum of natural history. But examples are still sometimes found in antique dealers' shops, in their original frames.

Some artists could not resist the temptation offered so generously by these sports of nature and set about turning them into complete and complex pictures by adding trees, animals, and human beings. Baltrusaïtis reproduces a large number of compositions in which artists like Johann König, Matieu Dubus, Antonio Carracci, and others pressed into service the bands and luminous depths of agate or the twists and turns of alabaster. When treating religious subjects, the painter drew from the stones' inner, translucent clouds a mysterious, supernatural, milky brightness: if he was depicting the crossing of the Red Sea, all he had to do was add terrified victims to the backwash of the dreadful wave suggested by the veins in the stone itself.

But it must be admitted that agate, alabaster, and jasper do not generally give the artist much to go on. In most examples the stone provides only a monotonous motif that cannot be used to portray much else but the curves of the clouds or the billows of the sea. The *paesine*, with their suggestions of cities or rocks, are quite another matter. Here the painter may supplement the buildings hinted at by the stone itself with little black rectangles representing doors and windows. He may also introduce trees, though he need not put in grass or undergrowth: the fine dendrites of manganese provide an excellent illusion of these. Human figures, of course, are due to the artist alone.

Two contrasting techniques are used. The first consists of filling the space available – the "sky" in the case of a ruin marble – with a single large figure. Baltrusaïtis cites for example a specimen in the catalog of the Settala Museum in Milan in which St. Jerome and the lion are depicted by the artist in a cave supplied by the stone itself. I myself possess a likely parallel to this in a stone, probably dating from the middle of the seventeenth century, which represents St. Francis praying to a crucifix while a death's head looks on from a rock. Only the saint, the cross, and the skull have been added by the artist. All the rest is the work of nature.

The second of the two techniques involves a closer alliance between the skill of the painter and the fantasies of geology and involves a number of figures and other more minor additions. In extreme cases the artist scatters them about so freely that it is difficult to distinguish the figments of his imagination from the contributions of the mineral itself. This type is represented by a cleverly touched up *paesina* in the National Museum in Havanna, where it is incorrectly catalogued as a painting on *agate* of the sixteenth-century German school. A huge disheveled pine tree, bereft of most of its main branches, stands on a steep rock where a donkey is grazing. The tree divides the scene into two halves contrastingly treated. The painter has made a considerable contribution to the right half, which shows luxuriant and varied foliage and other vegetation, together with two small figures engaged in animated conversation. The left half, on the other hand, has scarcely been modified at all, except for a square tower flanked by a couple of dark and slender yews. There is an urban panorama, made up of the broken lines characteristic of ruin marbles: the artist has done no more than pick out the bright edge of terraces and the dark curves of arcades. In the foreground some cliffs, by a natural symmetry, are reflected in a calm expanse of water; the sky above them and the river lapping at their feet are supplied by the ample curves which are almost always found in the mauve or turquoise part of this kind of marble. These undulations contrast with the straight verticals and obliques in the dark brown section of the stone, which suggest houses and palaces.

The finest examples of such works, in my opinion, are those hidden away in a small and little-known museum in Florence, aptly called the Opificio delle Pietre Dure, for it is at least as much a workshop as a gallery. The exhibits in question are small pictures – *paesine* are rarely more than fifty or so centimeters long – painted with energy, precision, and skill, all of them belonging to the first half of the seventeenth century. Four of them illustrate episodes from Ariosto's *Orlando Furioso.* Another represents Mary Magdalen in the desert. The most skillful of all is inspired by Dante's *Divine Comedy* and shows Dante and Virgil visiting Hell: red flames stand out against a dark ochre background seamed with black arborescences and dotted with pale blurs – the naked bodies of the damned. There seems to be a clear case of complicity here between the subterranean levels of suffering; and the genesis of a stone that itself comes from the depths of the earth, roasted in the heat of some nonhuman furnace.

It is no longer a matter of a painter's whim exploiting a strange material. Instead we have an encounter between a subject and a medium which might be called a demonstration of that subject.

As well as these six *paesine,* room 4 of the Opificio contains a painting on lapis lazuli which depicts the Lyciam peasants being turned into frogs by Zeus for refusing to let Latona drink from the pool where they were cutting reeds. Against the intense blue background of the stone, the goddess watches impassively as the impious rustics undergo their metamorphosis.

In such cases the painter cooperates with nature. Whether by just cutting out and framing the picture or by exploiting the pattern in the stone, he implicitly admits that nature, with or without an artist's collabouration, can produce arrangements of shapes and colours that may be accepted as works of art. In other words, it is as if nature not only provided a stock of models but also directly created works worthy of admiration – works capable of competing on equal terms with human achievements without having to pass through the alchemy of human art.

The interest in Florence marbles is now practically a thing of the past. What used to make them attractive was not their intrinsic aesthetic qualities but their resemblance to certain aspects of the external world, the closeness of this likeness or their apparent representation of some model. This last, this illusion of reproducing the manner of a painter, would be their chief attraction nowadays. But in the past it was as curiosities, as strange and marvelous objects, that they were prized. Even when they were polished and framed and treated as pictures, they were still sports of nature without any artistic value as such. The collector was impressed by a fortuitous parallel – the result of some obscure physical and chemical reactions – between the patterns in the stone and a vague vision of half-ruined walls and towers. Similarly, in agates the connoisseur looked for specimens with veins suggesting the manysided outlines of contemporary fortifications. In short, what he sought was an analogy, not a masterpiece of plastic art. This fact eluded people because in those days it was impossible to conceive of a painting that did not reproduce some object. Pascal could condemn the whole art: of painting by writing: "What a vain thing is painting, which evokes admiration through a resemblance to things of which we do not admire the originals."[8]

A modern connoisseur, however, is more likely to be attracted by ruin or landscape stones in which the alleged resemblance is imperfect or rudimentary: photographic likeness merely takes him aback. It is what may be called the style of the stone that interests him: the simplifications, the broken lines that separate the different colours and seem to break up the light. A resemblance that used to be regarded as rough or imperfect now appears the deliberate result of some bold or inventive stroke or felicitous transposition. This reversal of taste reflects a general evolution in the attitude toward picture stones. But let us now examine some other curious examples in this field.

Dream Stones

In China in the mid-nineteenth century an artist might select a slab of marble with marks or veins that appealed to him and then proceed to trim and frame it, give it a title, and engave his seal on it. He thus took possession of the stone, turning it into a work of art for which he was henceforth responsible. I have published a study and reproduction of one such specimen; the Chinese called them dream stones.[9] The example in question is called *The Solitary Hero* and bears the seal of K'iao Chan. It shows a shape in yellow, sepia, and black which may just about be taken for a human being crouching against a whirlwind. I have since seen many other stones of this kind, and shall here consider two of the larger ones, representing landscapes. They show wooded ravines with lakes and pools reflecting nearby cliffs, or else with flat shores vanishing into the distance in one direction and in the other blocked by sloping hills apparently bent in the direction of the prevailing wind. These marbles, like my first example, are signed, but instead of a title each one bears a little poem in keeping with the scene, wild or peaceful, that it depicts. The first poem runs as follows:

Amid the greymost mountains
The clouds come and go
Like so many great men:
None reaches the heights,
For that is too difficult.

The texts which accompany the reproduction form a whole, on the theme of nature as a painter. The second:

In its ring of mountains
An ancient temple looks new.
Here the spirit seems as keen
As the wind, the air, the trees.
How can such a feeling be expressed?

The description of a luxurious Chinese interior in a novel by Lin-Yu Tang shows that the love of dream stones was widespread:
Everything was grandiose, simple and severe. Against the Northern wall stood a high mahogany table, massive yet sober in design, on which there was nothing but three objets d'art. In the middle was an ancient three-legged vessel in cloisonné inlaid with gold. Then came a slab of rare marble sixty centimeters square, with natural pictures suggesting a landscape enveloped in rain and mist, with almost invisible mountain peaks, woods, and two fishing junks, all incredibly lifelike. The natural patterns on the other slab of marble suggested a large duck: the head, beak, and neck were almost perfect, while faint lines seemed to depict the body, and brown marks the bird's webbed feet.[10]

Admittedly resemblance is still the criterion, but now the preferred likeness is evasive, allusive, even ambiguous. At all events, the figurative factor has given way to a plastic quality. Though I have sometimes put the theory forward, on reflection I am not sure that the artist only affixed his seal to a rectangle of marble selected by himself: I wonder whether he did not in fact allow himself to add imperceptible improvements to the natural picture.
He may have used some unknown but easily imaginable process that so to speak "mineralized" the colouring agent employed, making it take on the grain and polish of the stone itself and become an integral part of it. I can think of no other way of accounting for the fact that the veins and marks

and motifs in the Chinese marbles are visible on one side only of the slab, whereas normally they would go right through the stone, varying somewhat on the two faces, as in ruin and landscape marbles and agates. Whether or not some subterfuge is involved, the point is that unlike the Western artist, starting from a recognizable resemblance in the stone and emphasizing it in order to change it into a picture, the Chinese painter uses his art to put the connoisseur on a false scent, trying to make him think the picture a natural, untouched up marble which the artist has only given a title, that is, merely interpreted and signed.

The two methods are diametrically opposed, but they both reveal significant connections between art and aesthetics, if I may call art the beauty deliberately produced by man and aesthetics the appreciation of all beauty, both what derives from art and what is met with accidentally in the universe.

Language does not distinguish between the two orders of beauty. We speak indiscriminately of a beautiful cloud and a beautiful statue, a beautiful tree and a beautiful line of poetry, a beautiful landscape and a beautiful picture; and while the last may represent a beautiful landscape, it is not considered beautiful for that reason. Yet what is admired in each domain is, or appears, fundamentally different (the nuance hardly matters, for here being and seeming tend to coincide.) And when we bracket the two realms together we confuse the results of two unrelated if not contradictory methods. The fact that the beauty created by the artist imitates or reproduces what is found in nature makes possible and even encourages a common vocabulary.

In theory this situation ought to change when painting deliberately, even systematically, turns aside from the forms offered by the external world, and the painter takes care not to set down on canvas anything that may be recognized or identified. In these circumstances, if I wish to find some natural object, some stone that may be compared to a picture, I shall no longer look for a possible resemblance but for an equivalent of the effects which are now being sought after and which are entirely plastic and without reference to the forms in the visible universe. To juxtapose with works representing the new style of art I shall choose a new kind of mineral, bearing scars equally far removed from both geometry and meaning and with lines less linear than the veins in marble, less circular than the whorls in agate. These messageless hieroglyphs for the most part evoke no memory of any appearance. I refer to the stones known as *septaria.*

Septaria

Septaria are siliceous nodules crisscrossed with cracks filled with calcite. They appear to be little known, though there seems no reason to suppose they are rare. Septaria are quite varied, since the name applies not to a particular species of mineral but to a structure which may occur in several. The most common ones on the market come from Spain and the Dakotas, and it is these I prefer to concentrate on. English septaria show rough, even confused patterns, sometimes very large, and are either round or long and almost rectangular. In the last case the markings are independent of one another and look like runes. Septaria from Saint-Wendel in Germany have the most delicate markings, with the finest lines; they are often outlined in pink. From northern Bohemia, where many small agates like jasper in appearance and colour are found, I have been sent a specimen with the quartering structure characteristic of septaria: the markings resemble a chrysanthemum whose short, rosy, flamelike petals edged with blue writhe against a background of white. At the other extreme, among the soft rocks, similar motifs occur in the large cracks to be found in certain septated marls. But whatever their provenance and composition, septaria never exhibit the same pattern twice, and it would be impossible to divide their markings up into categories. One nodule alone contains potentially many and various motifs, depending on the different angles at which the stone might be cut. The cross section taken of the nodule, which is a kind of elongated and flattened disk, may be made lengthwise, crosswise, diagonally, or equatorially, and the choice determines what aspect the inner maze of crevices will present. Even patterns situated on the same axis change so fast that at first sight there seems to be little in common between the two sides of one thin slab or the motifs in a couple of adjacent sections, though the eye can easily supply the transition for itself.

As for the style, the motifs may be simple or elabourate, ordered or tangled, symmetrical or chaotic, made up of thin lines or large patches; they may be all in one colour or range through every shade of sepia, ochre, milky white and, in the metallic parts, burnished steel. Sometimes the patterns emerge through the surface of the nodule, so that a ridge of sharp spikes prolongs the hidden pattern externally. The smallest and simplest septarium I have ever handled is about the size of an aspirin and has at its centre a pattern like a tiny four-pointed star, or rather the deep blue flower of certain dwarf mountain crucifers with its narrow, almost threadlike petals. Sometimes the tips of the petals are black, edged with the same white as that in the centre of the flower; rays growing ever thinner trace the sparest of images. But the pattern may become more complicated. The branches end in hooks or fingers, or are joined together with almost invisible rigging. Webs and nets start spreading over larger surfaces, some as fine as the hair cracks counted a defect in fine porcelain even though they can scarcely be seen. The lines are sometimes lighter and sometimes darker than the stone they run through. More rarely they are rust-coloured. Occasionally they open out into ravines lined with little crystals. They form patterns which explode; showers of manysided cells; sprays of dodecahedra all on one plane; irregular veins branching out in all directions then suddenly tapering away; steelyards weighing a large object which is yet so light that the arm of the balance is unmoved; cobwebs spun in the void, attached to no point and containing no lurking spider; cross sections of murexes, with the helix in the middle and the spines on the outside; the waving tentacles of sea anemones; the filaments of jellyfish, ending in a whiplash. Out of the dark of the stone, between the beams of an incandescent star, shine bright streaks and points like dandelion seeds blown from the stem, fixed in their flight and forming a kind of halo around their original centre. Such specimens always exhibit a geometry that is both capricious and harmonious, airily combining rigour and ease. The nucleus of the composition seems to be the centre of the nodule itself; the pattern looks as if it is fleeing wildly from that centre, then fading out and vanishing before it reaches the periphery of the stone. But it divides up the space unequally, according to an easily detected rhythm. The bold, flowing lines print on the unyielding silex the idea of, the recipe for, a freer and more complex motion than that of waves spreading outward from a disturbance in the surface of a liquid. What we have here is a sharp-angled craquelure making its way through a solid medium and soon brought to a halt by the strength of the stone. In some exceptionally large specimens the cracks seem to radiate from two adjacent centres. The wide-meshed webs approach one another without overlapping, explore one another without joining, brush like antennae, like anxious or voluptuous feelers. It is as if the two systems were really conscious organisms trying to get to know one another by cautious touches. They are still only diagrams, closer to the blueprint than to life, but their polarized substance seems to stir. They might be two ameobas about to fight or mate, but because they are frail, perhaps soluble, they feel the need to get acquainted first, though they fear the first contact may prove to be corrosive.

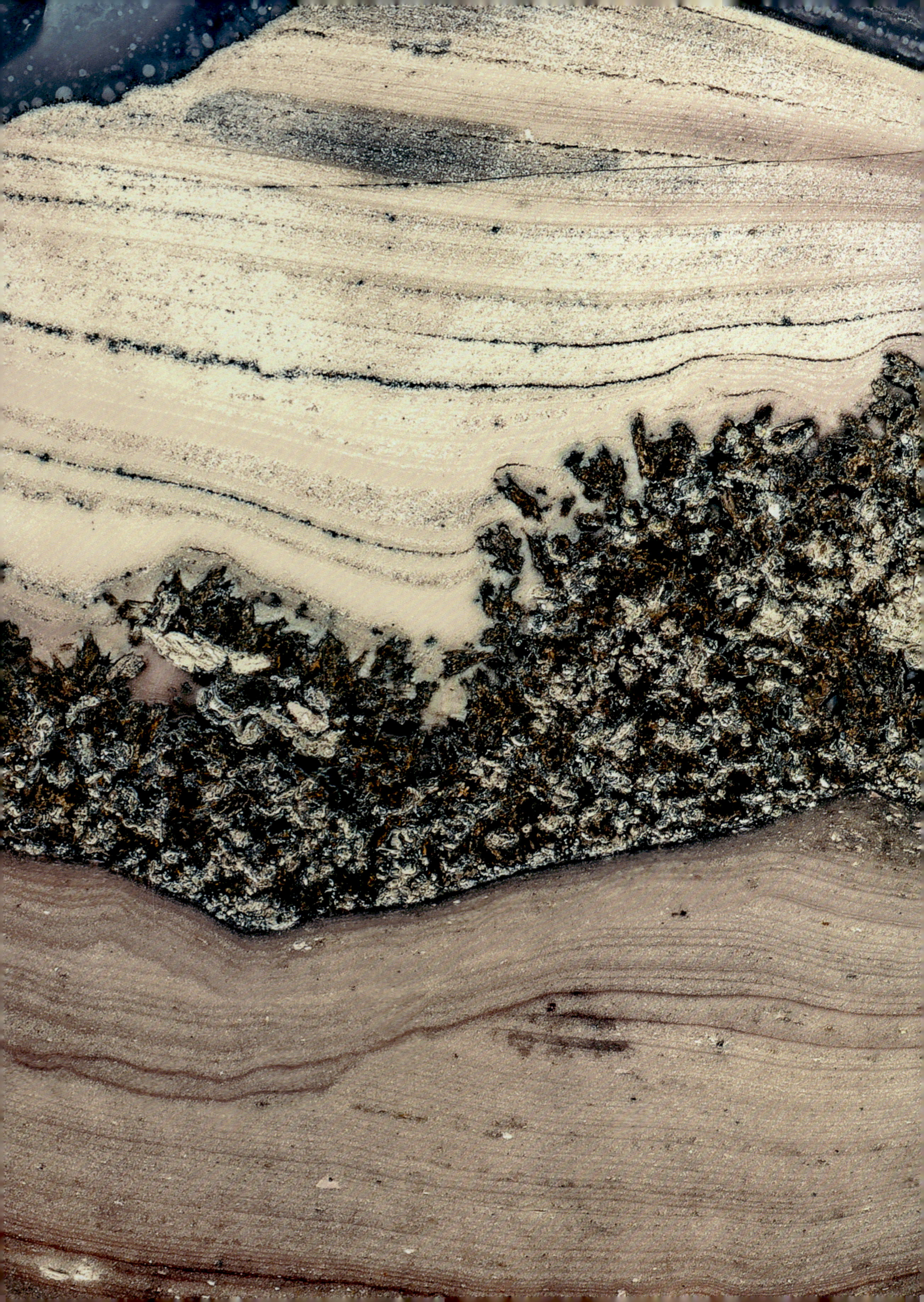

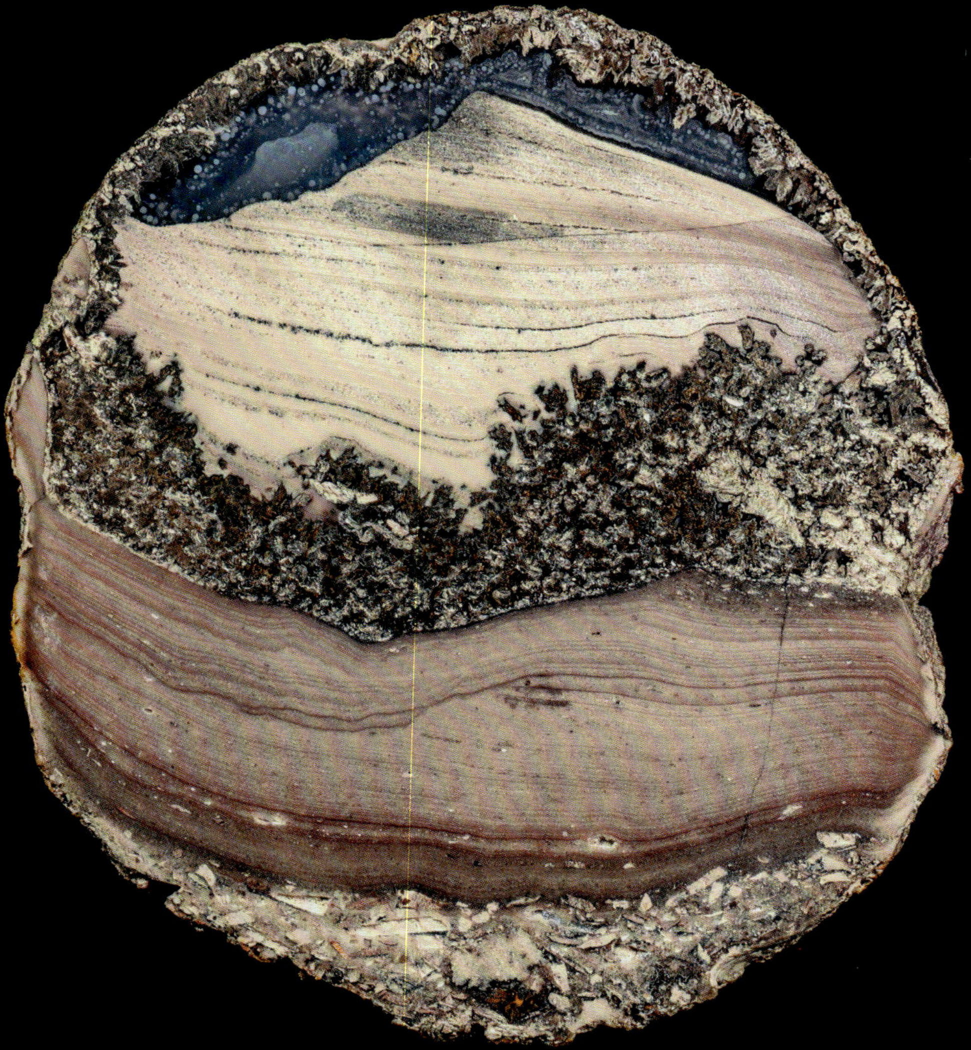

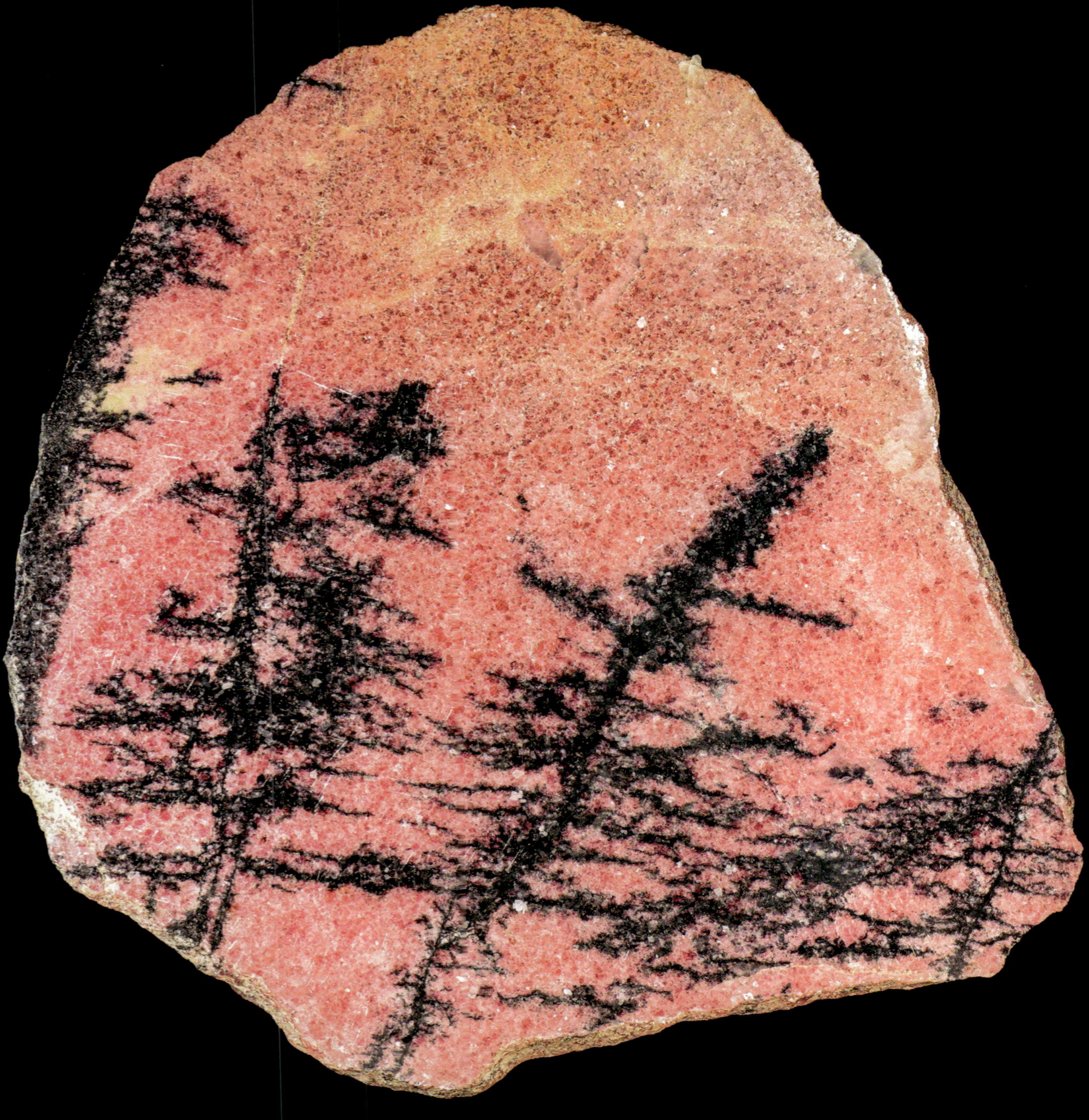

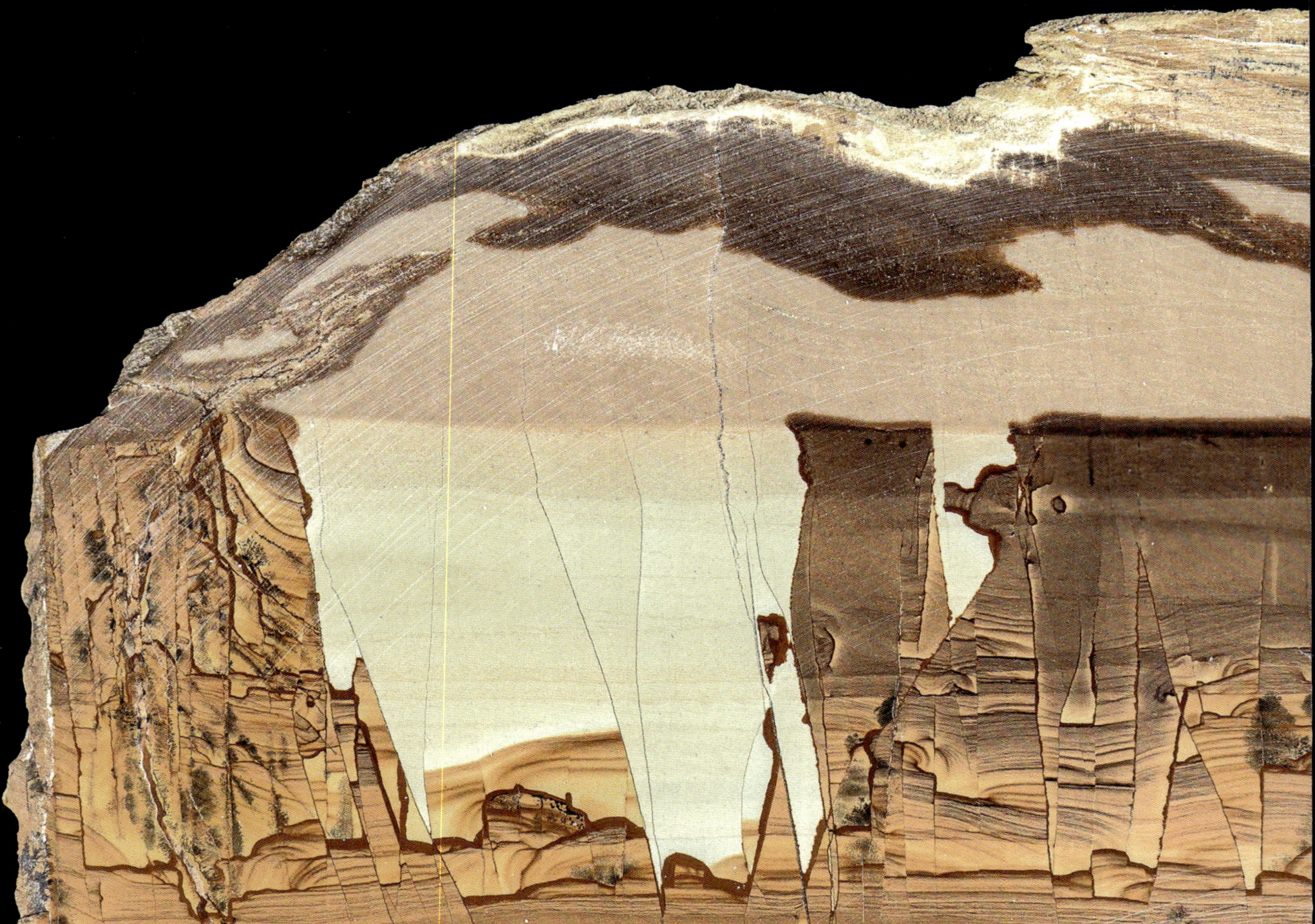

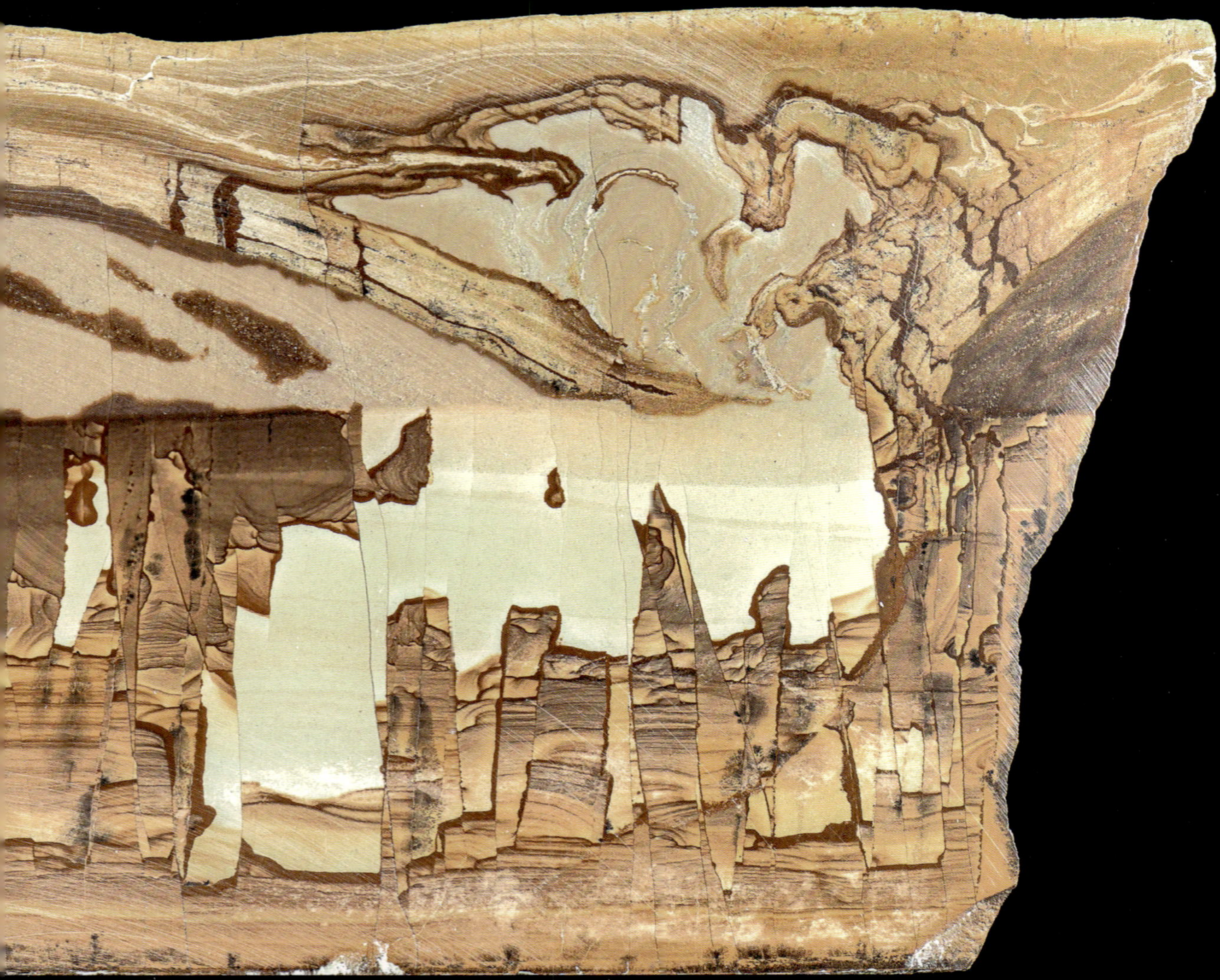

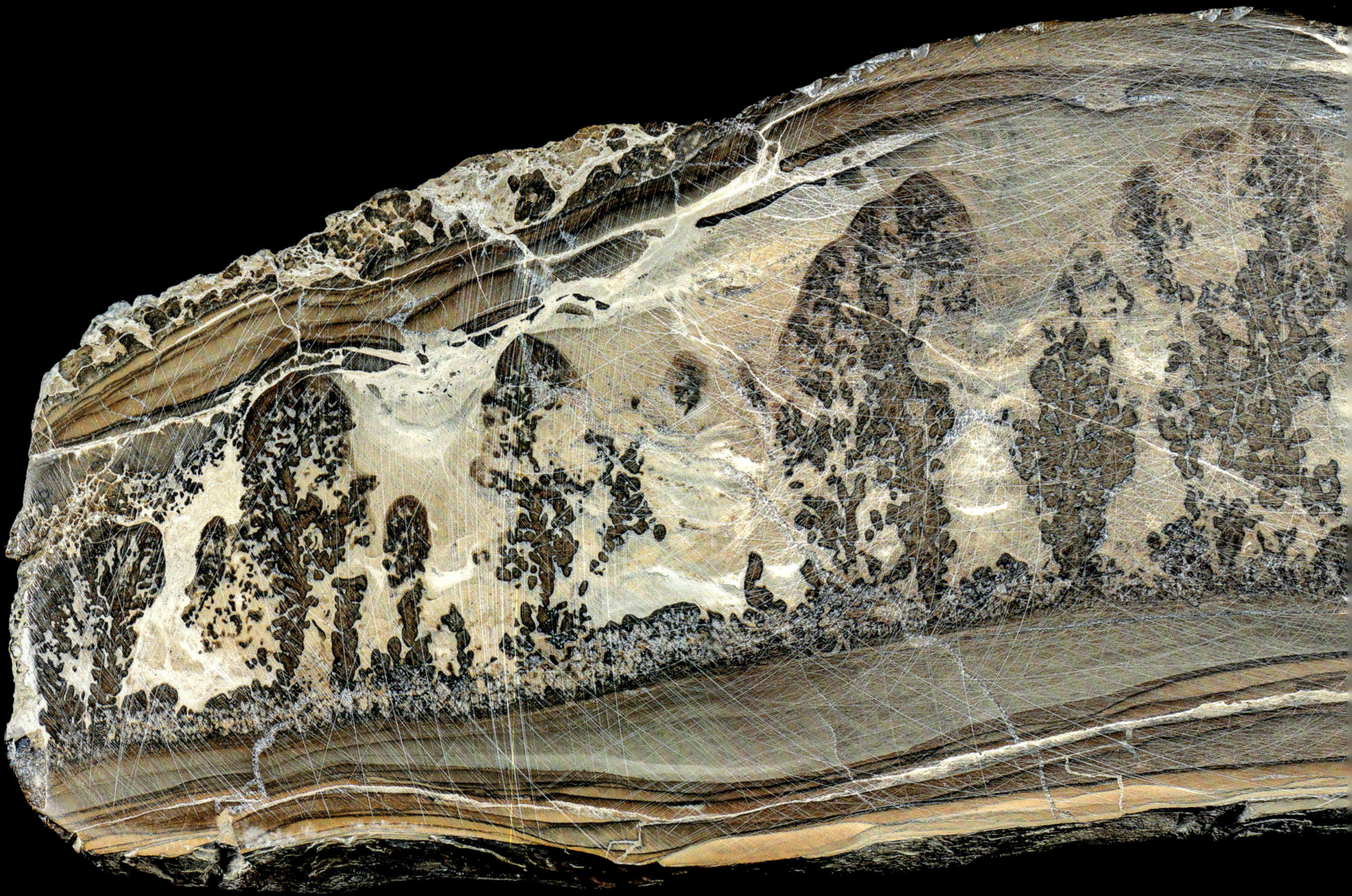

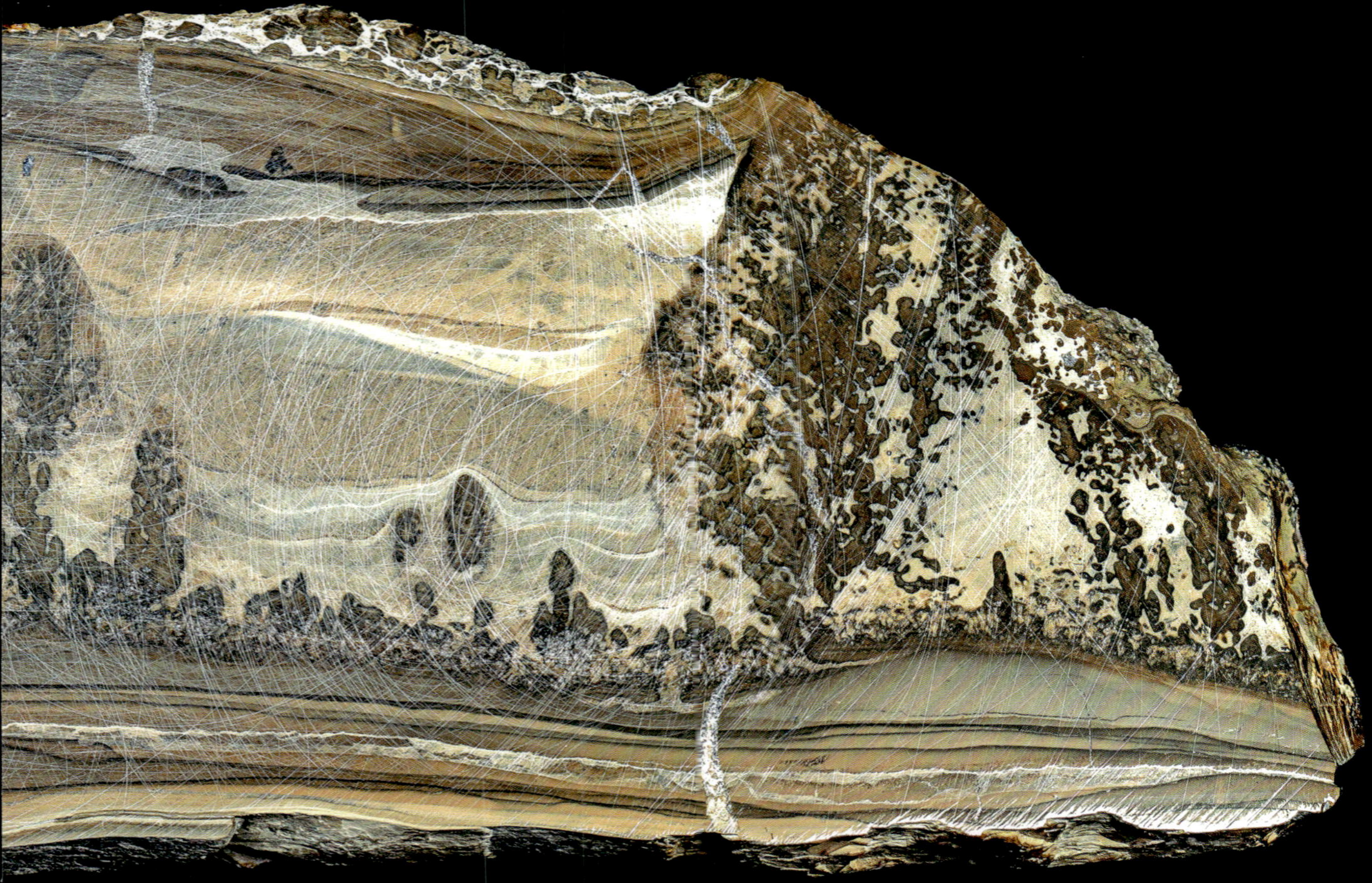

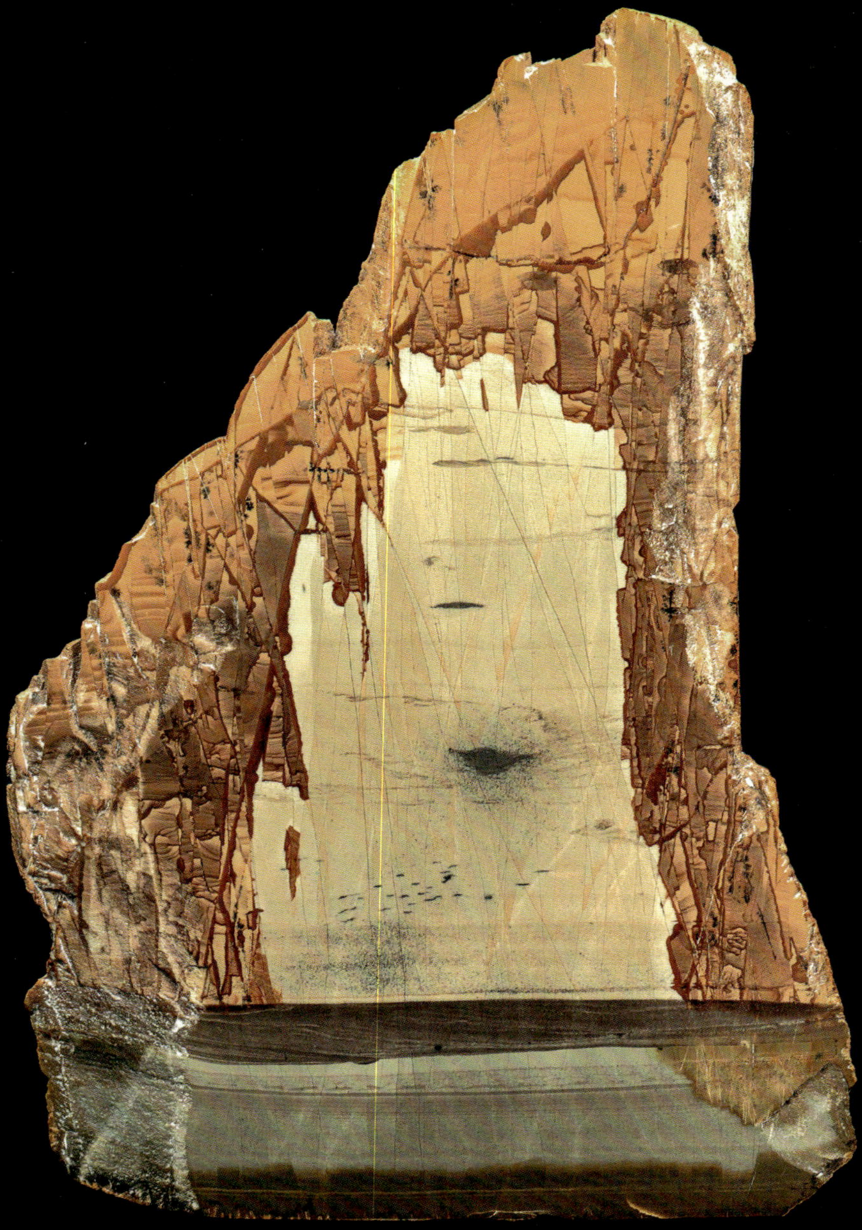

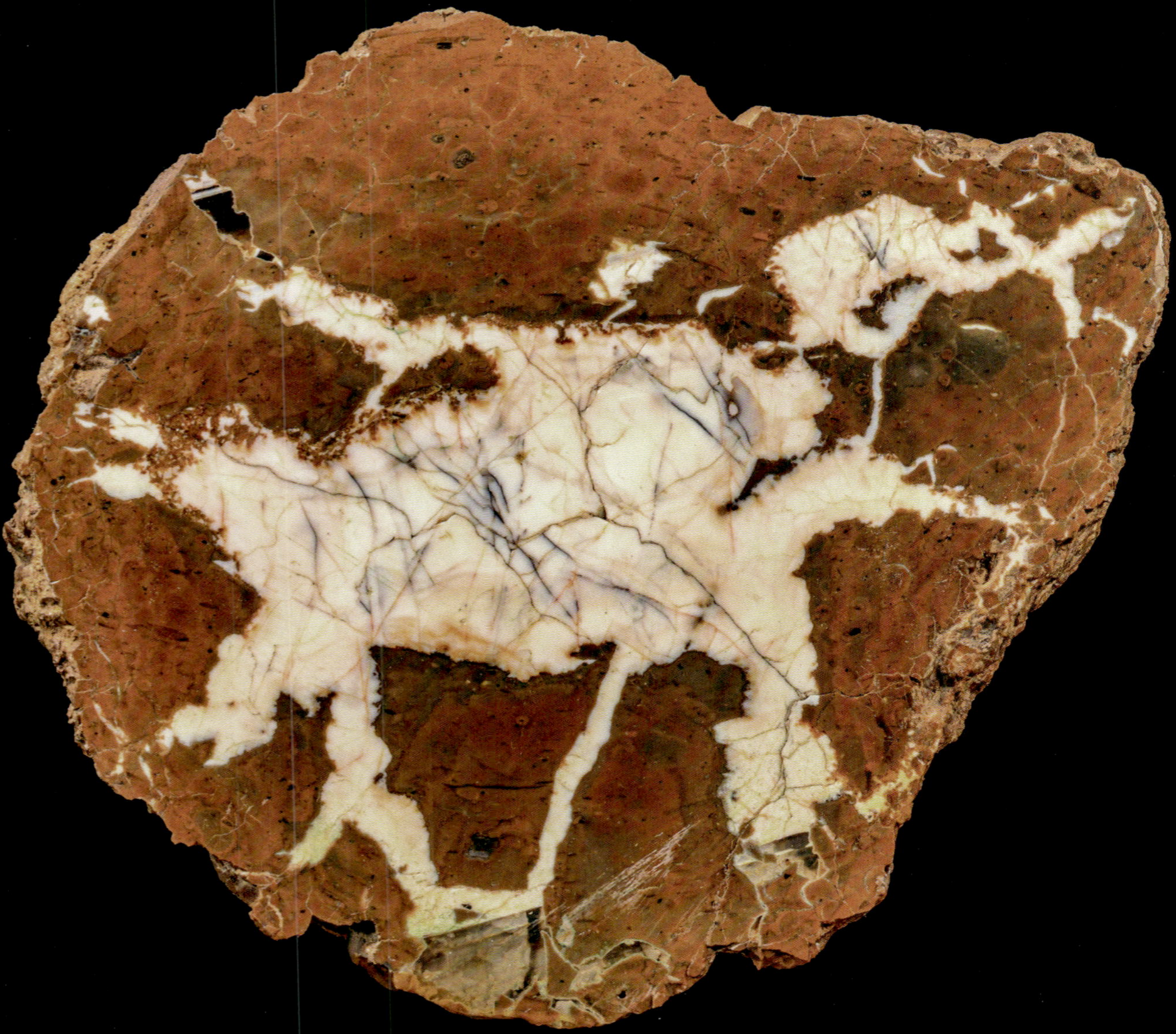

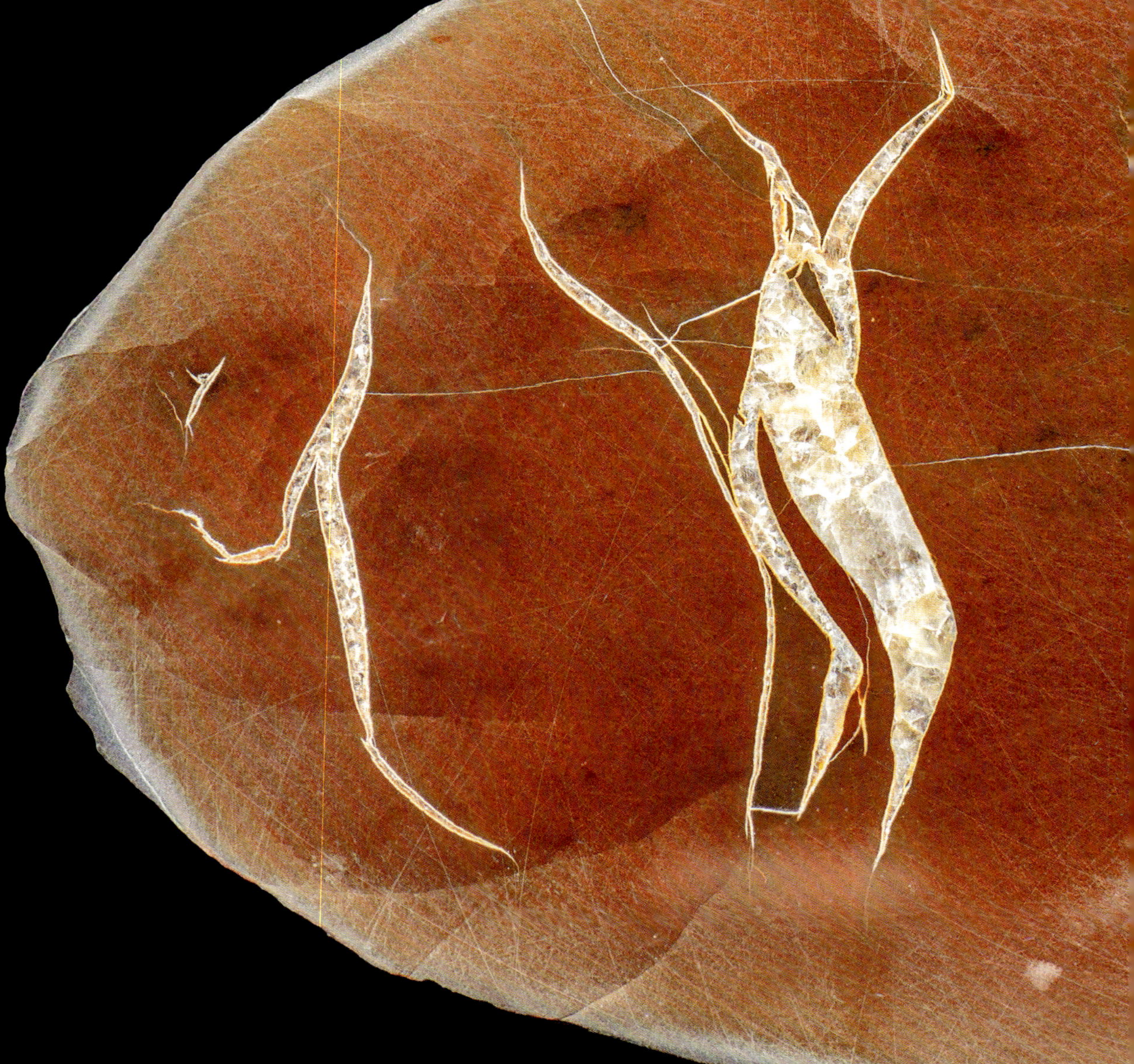

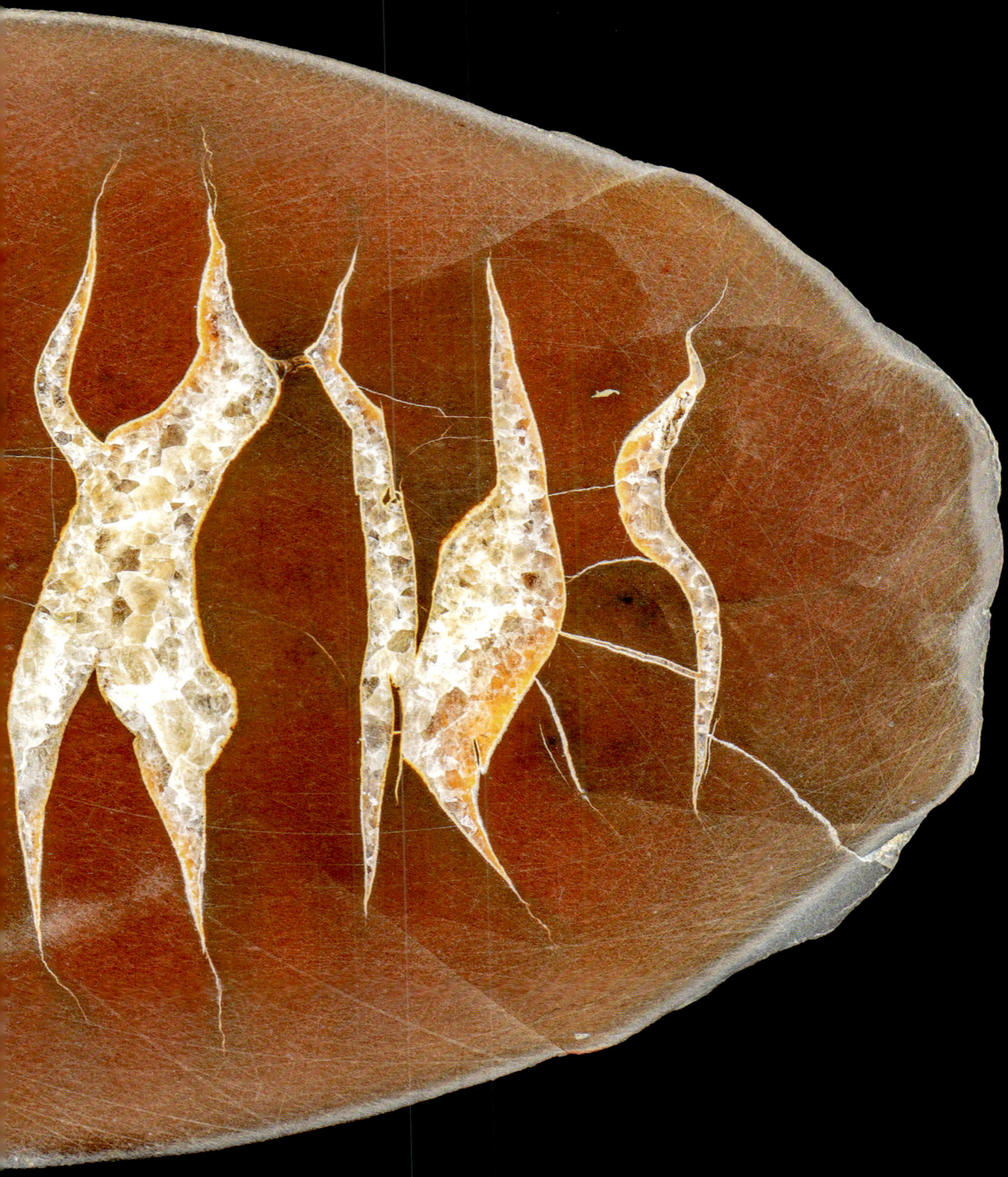

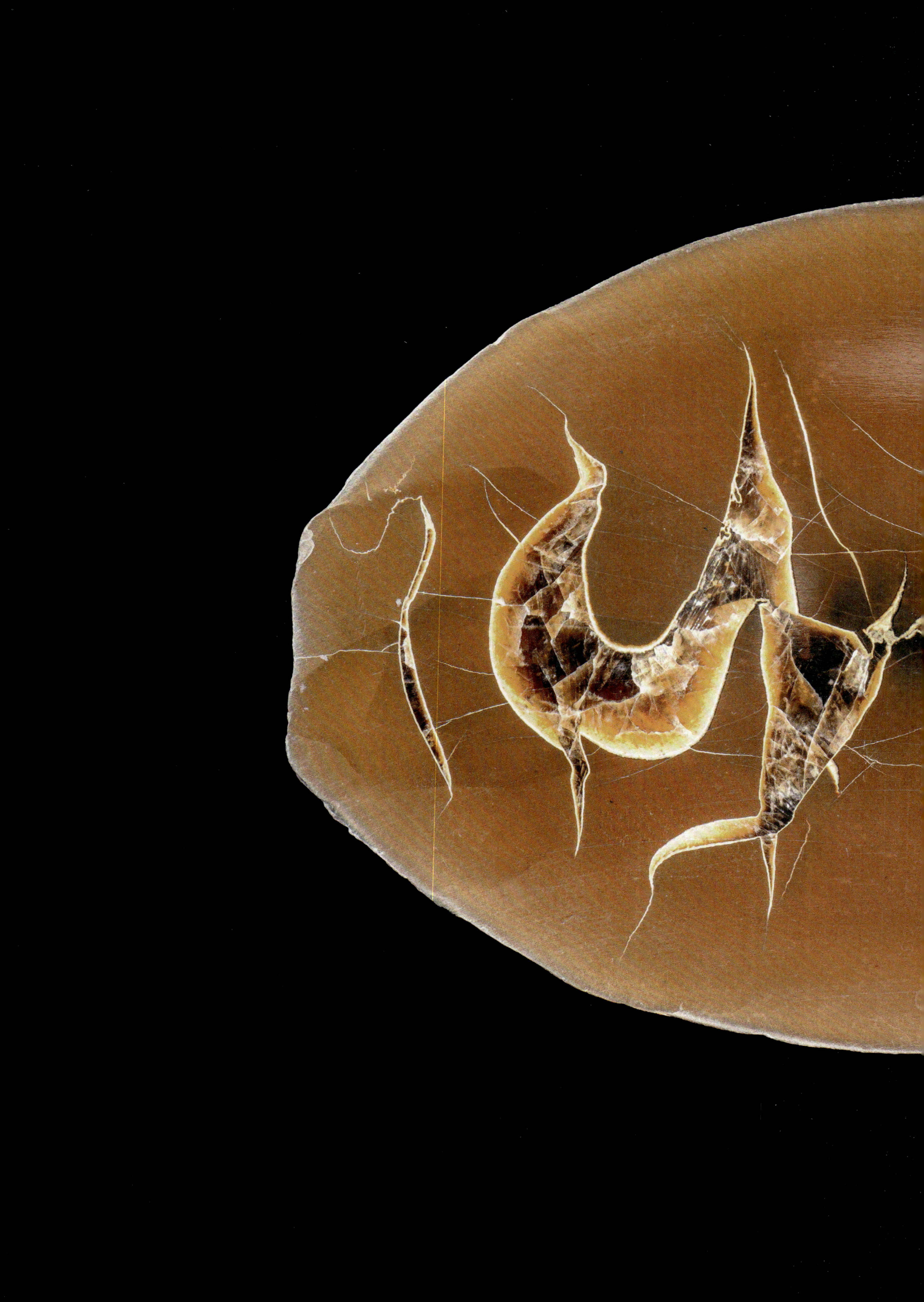

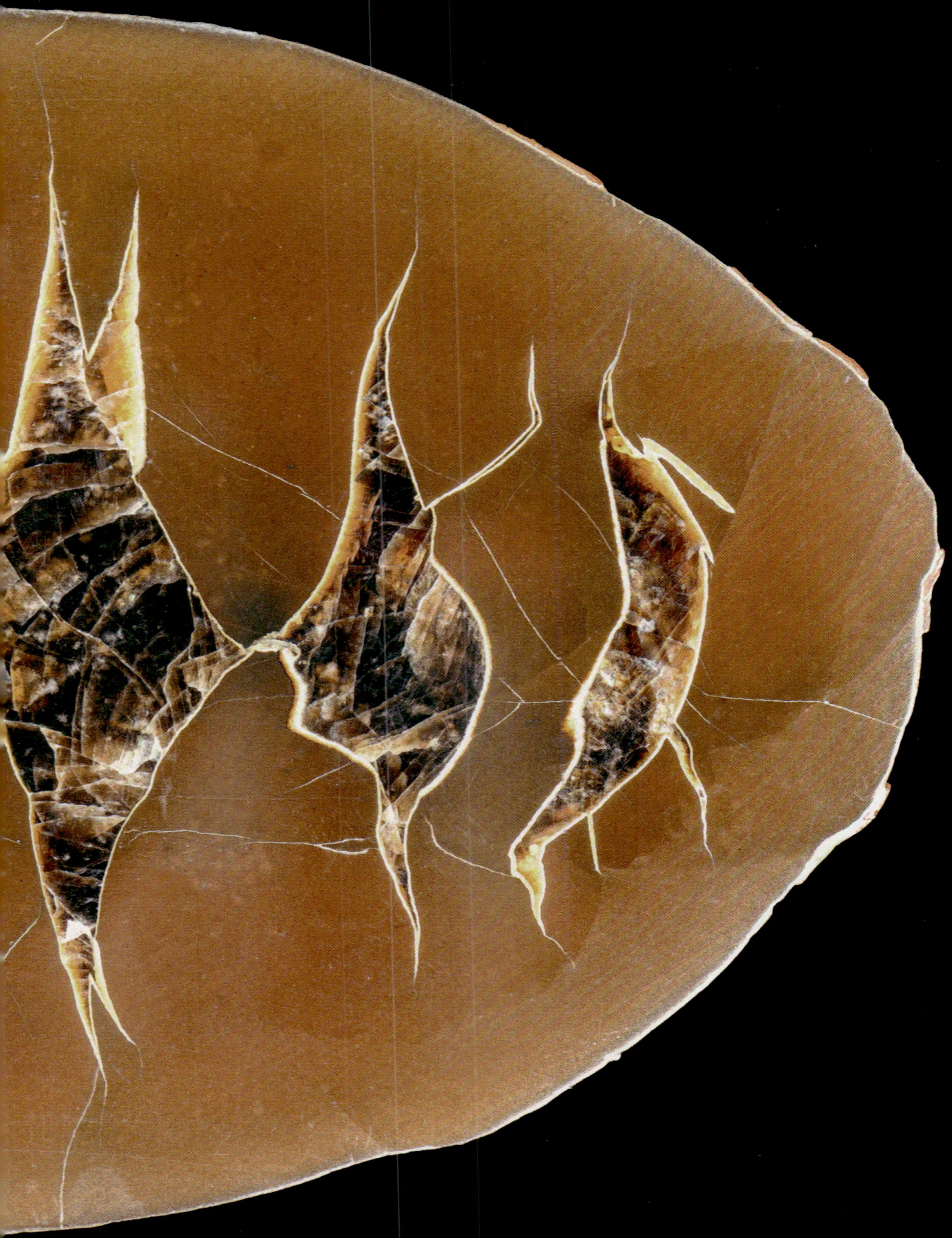

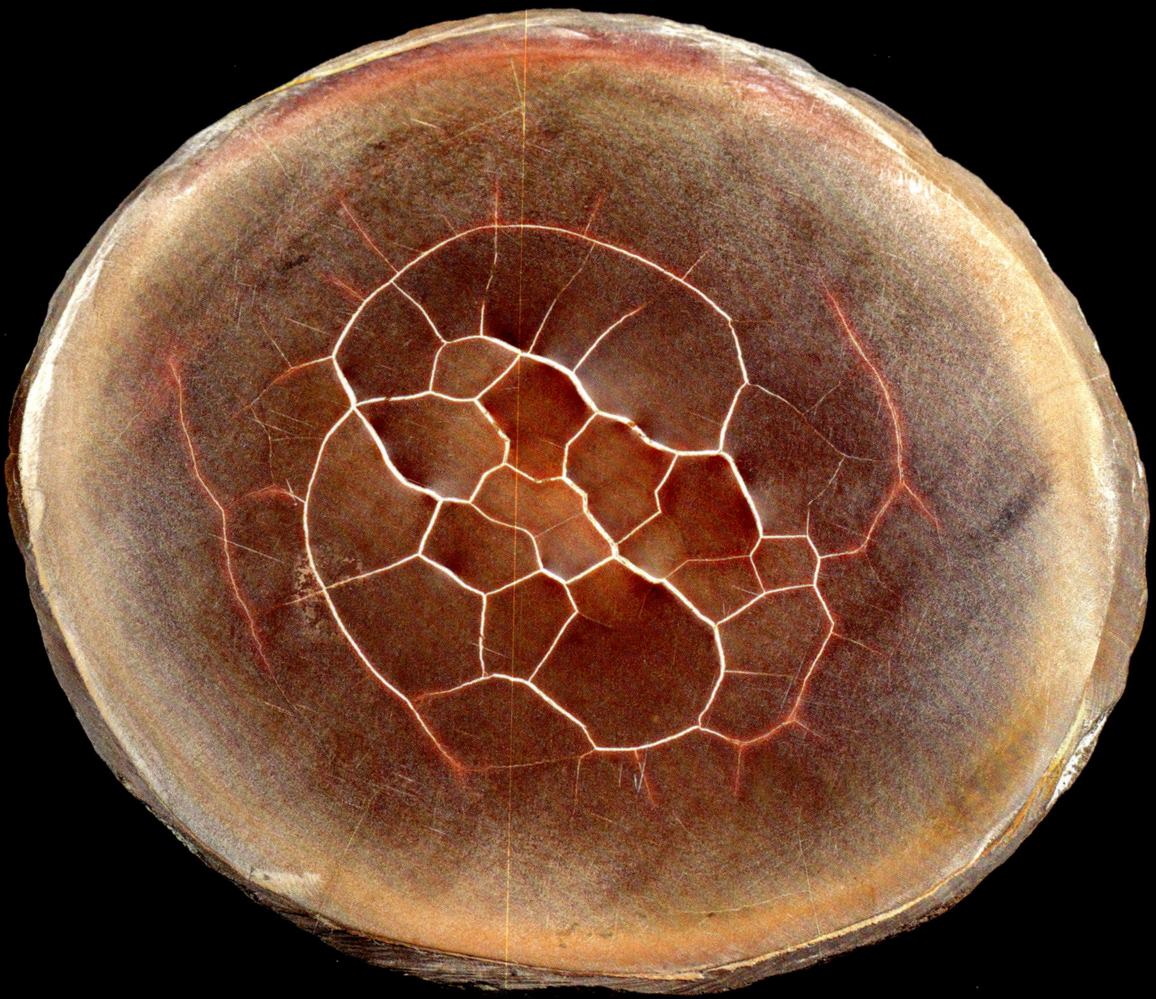

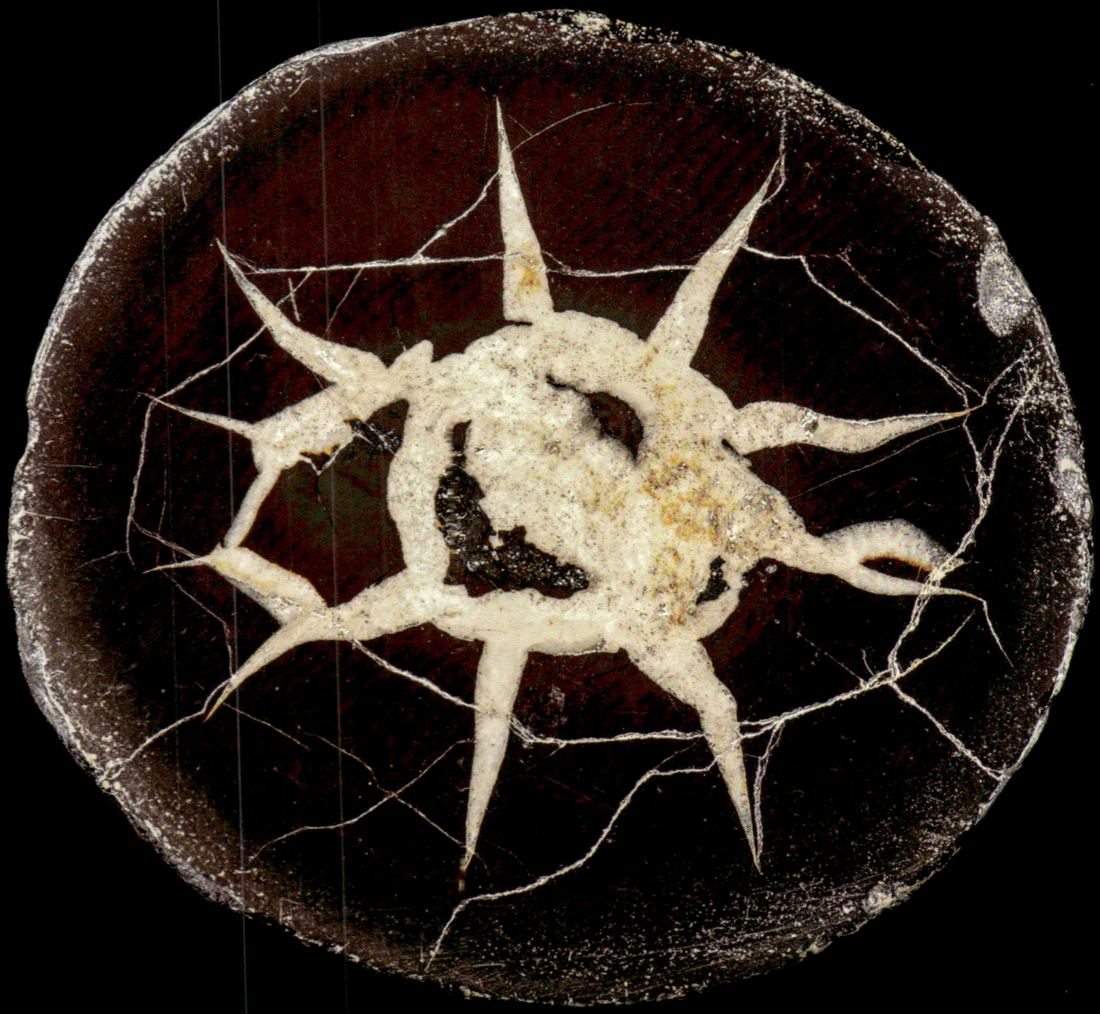

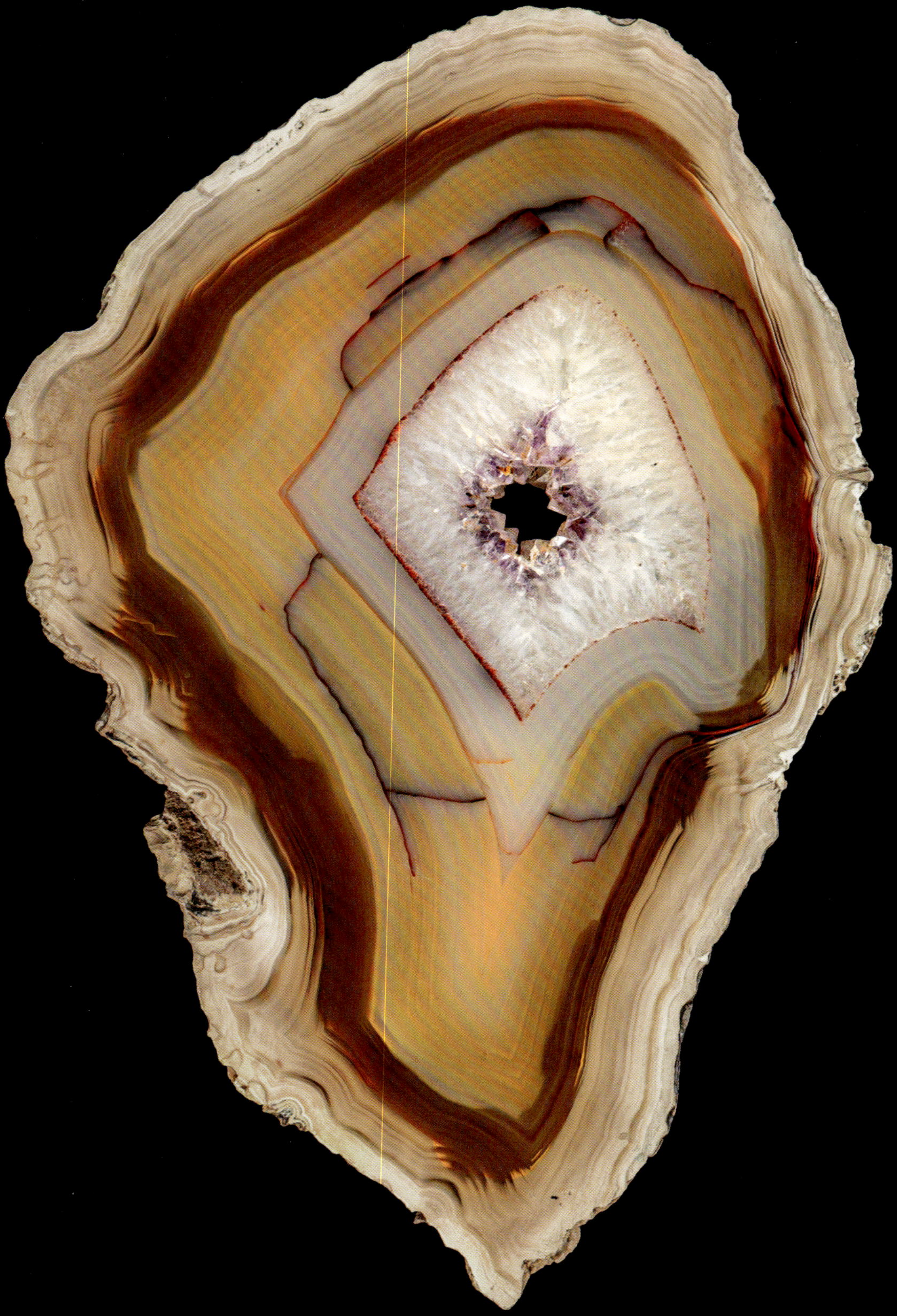

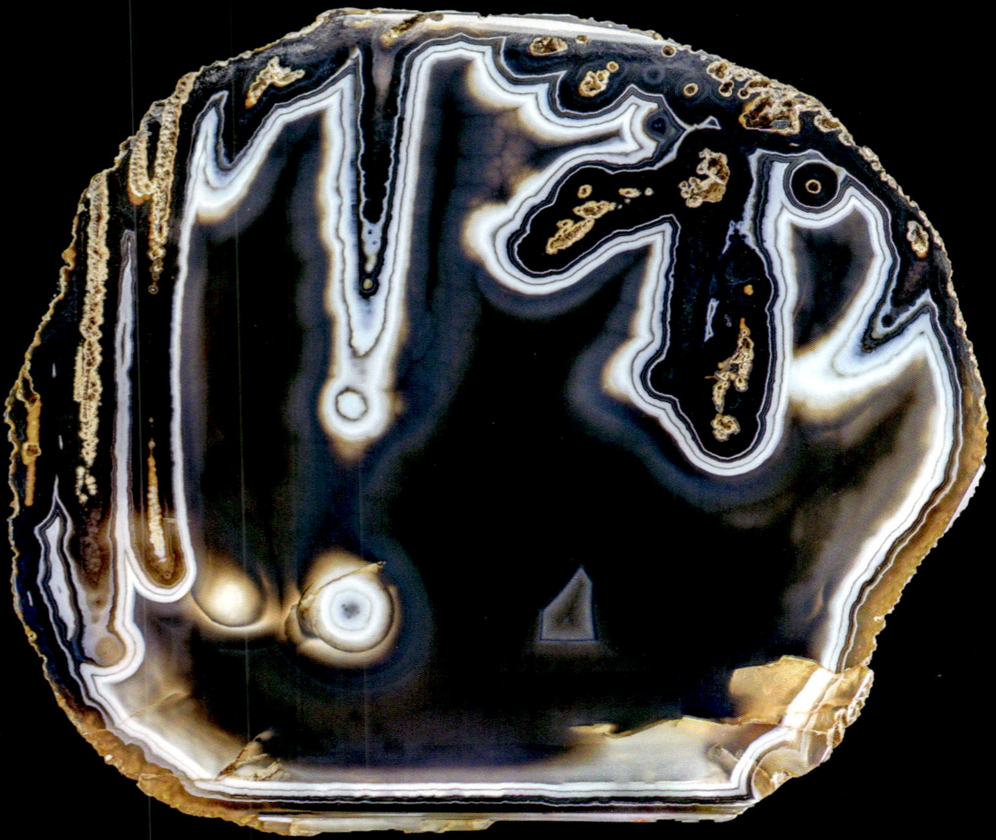

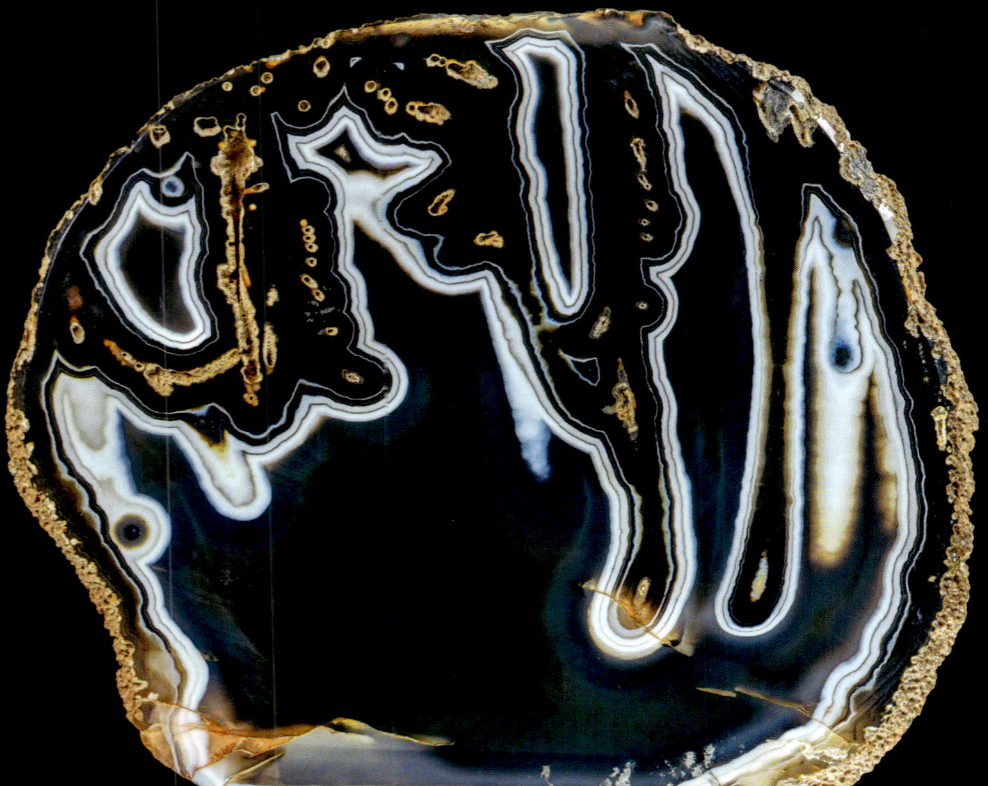

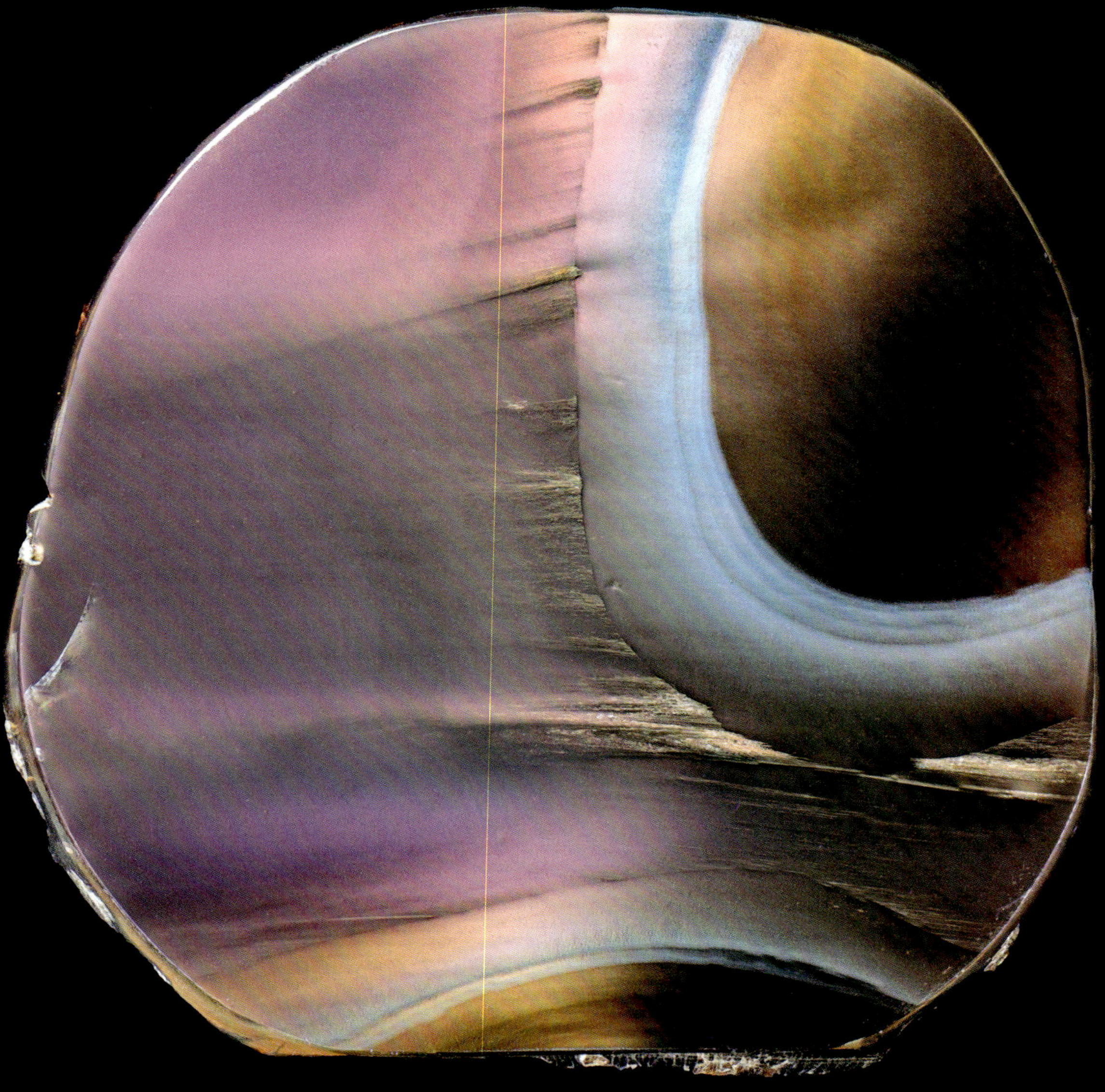

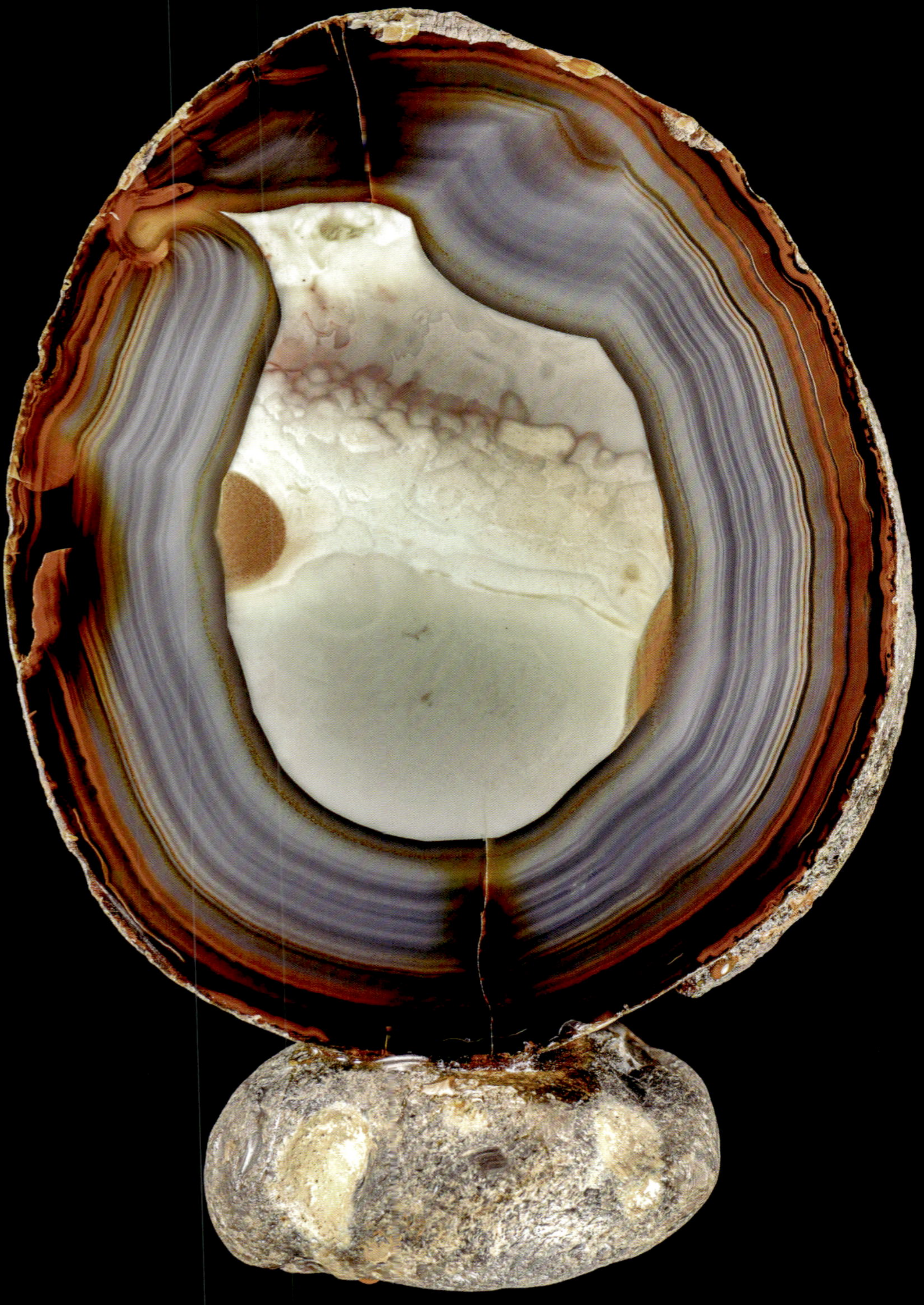

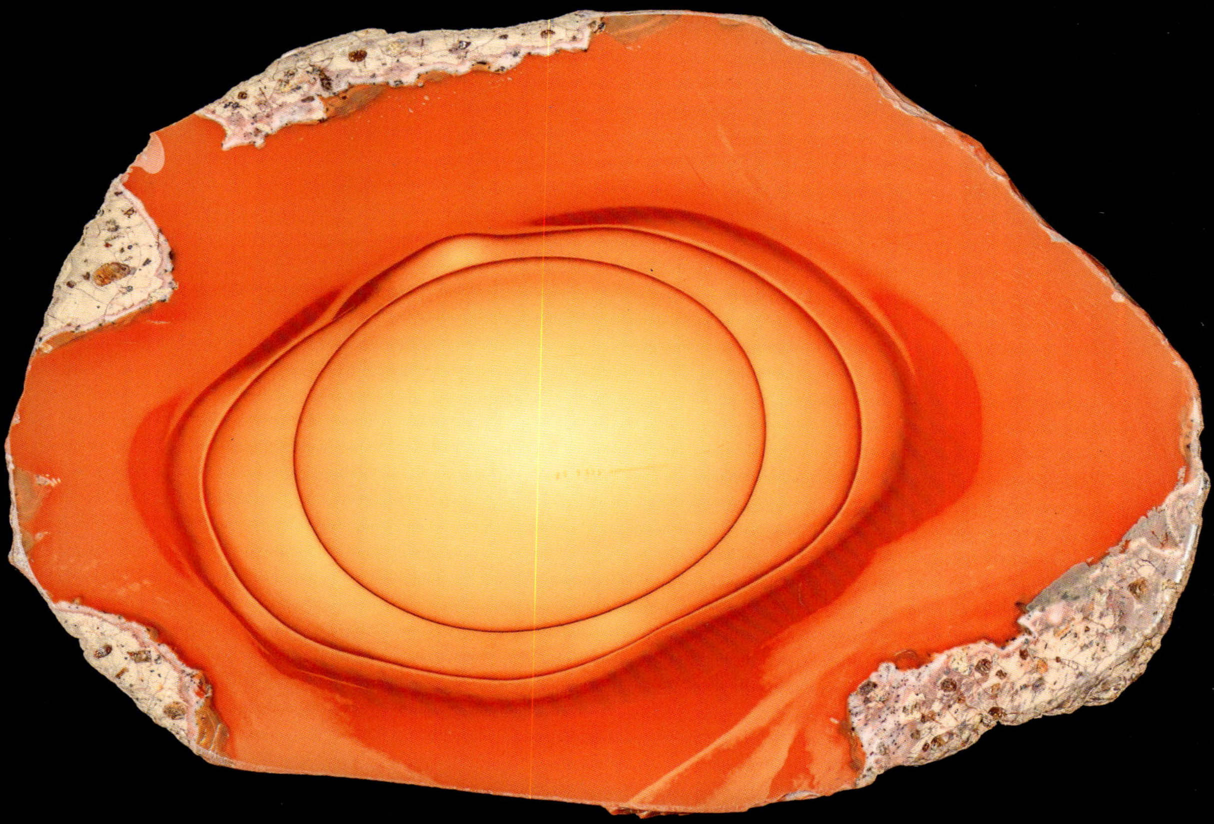

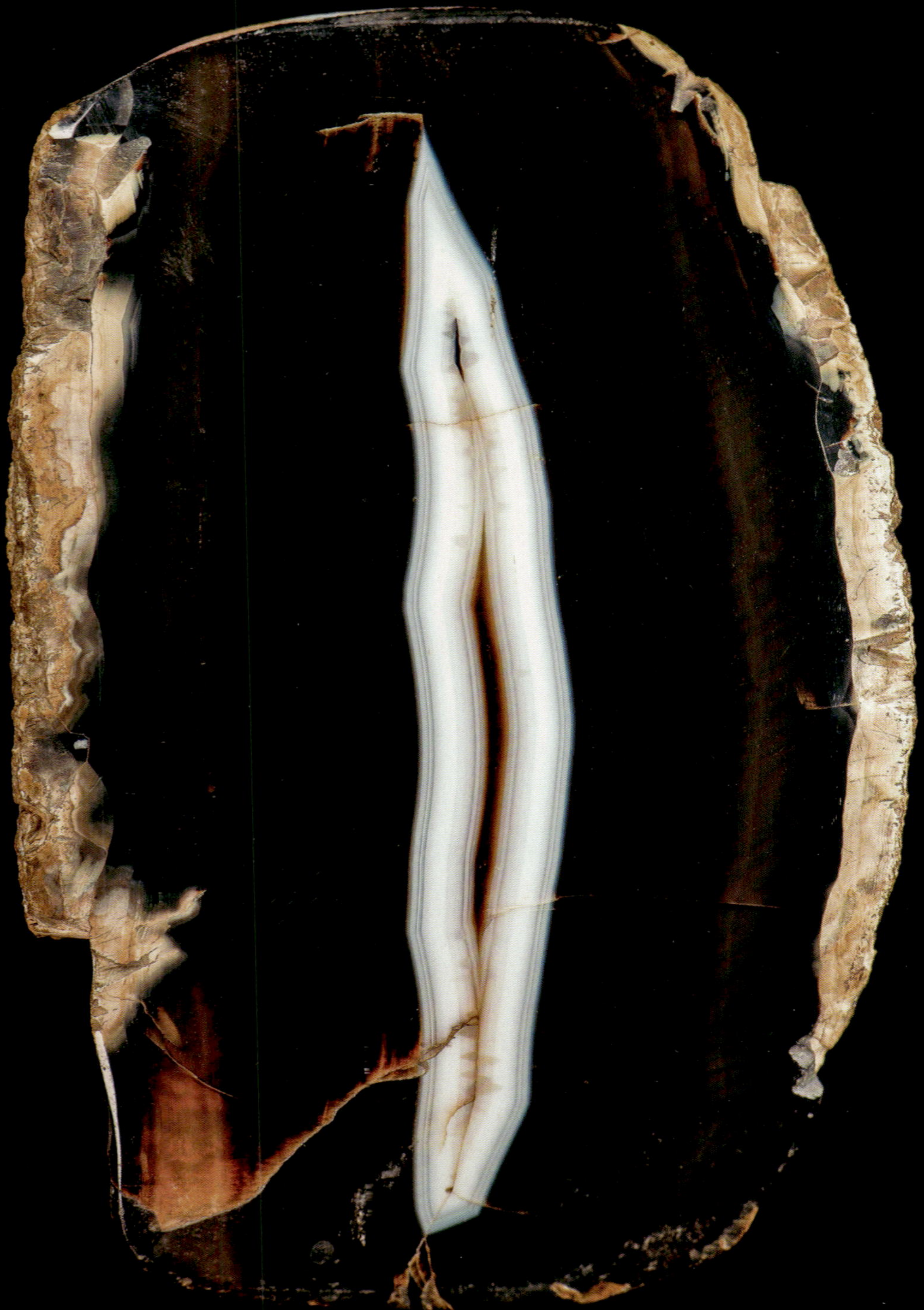

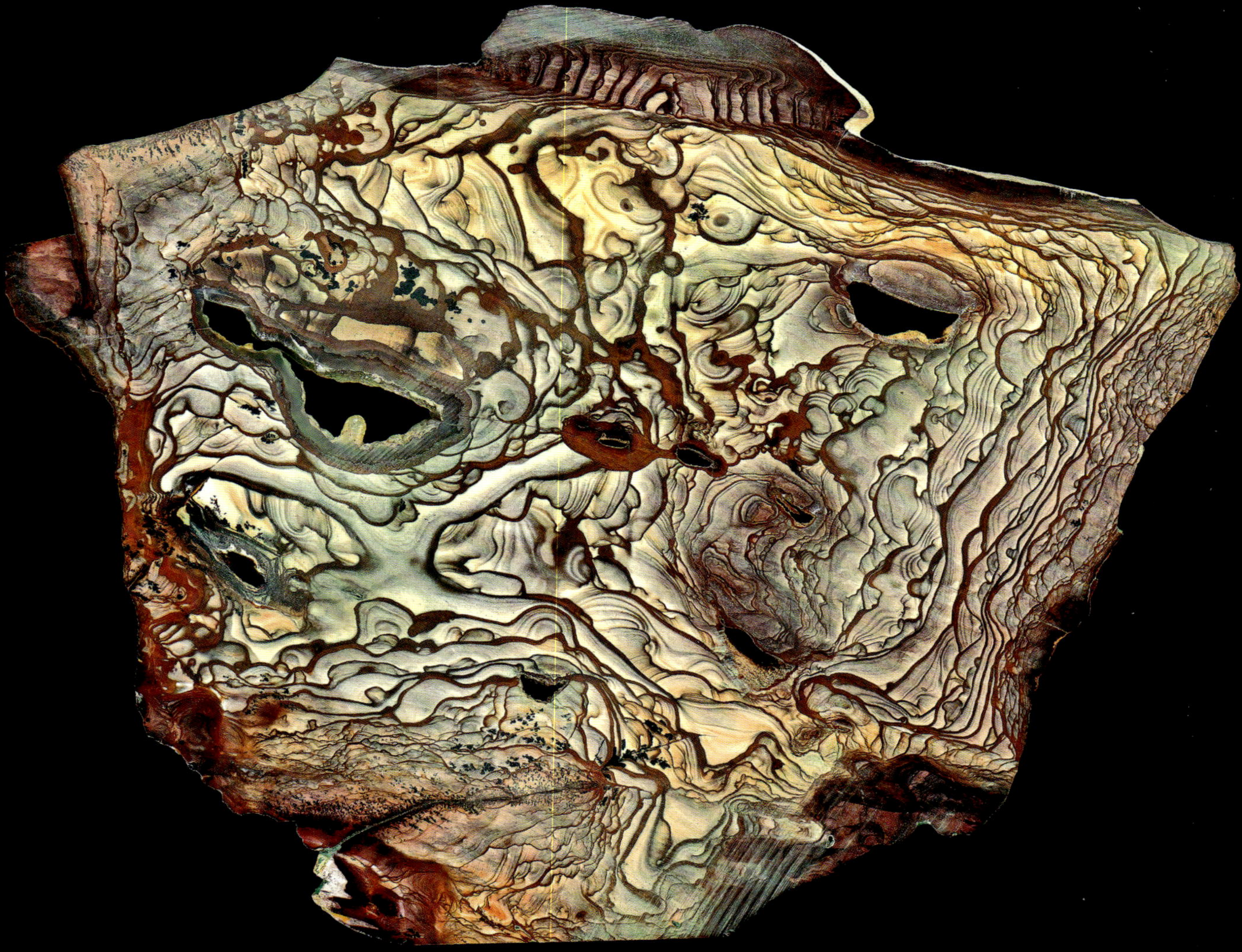

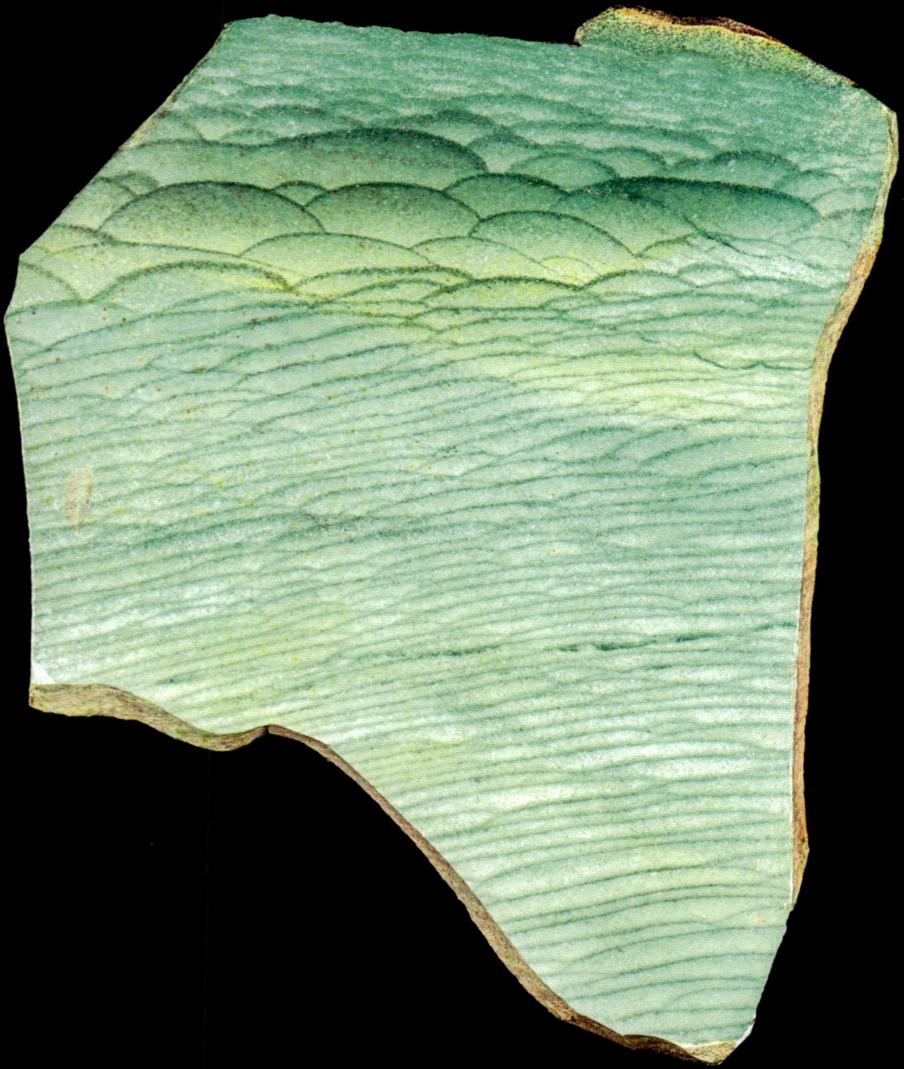

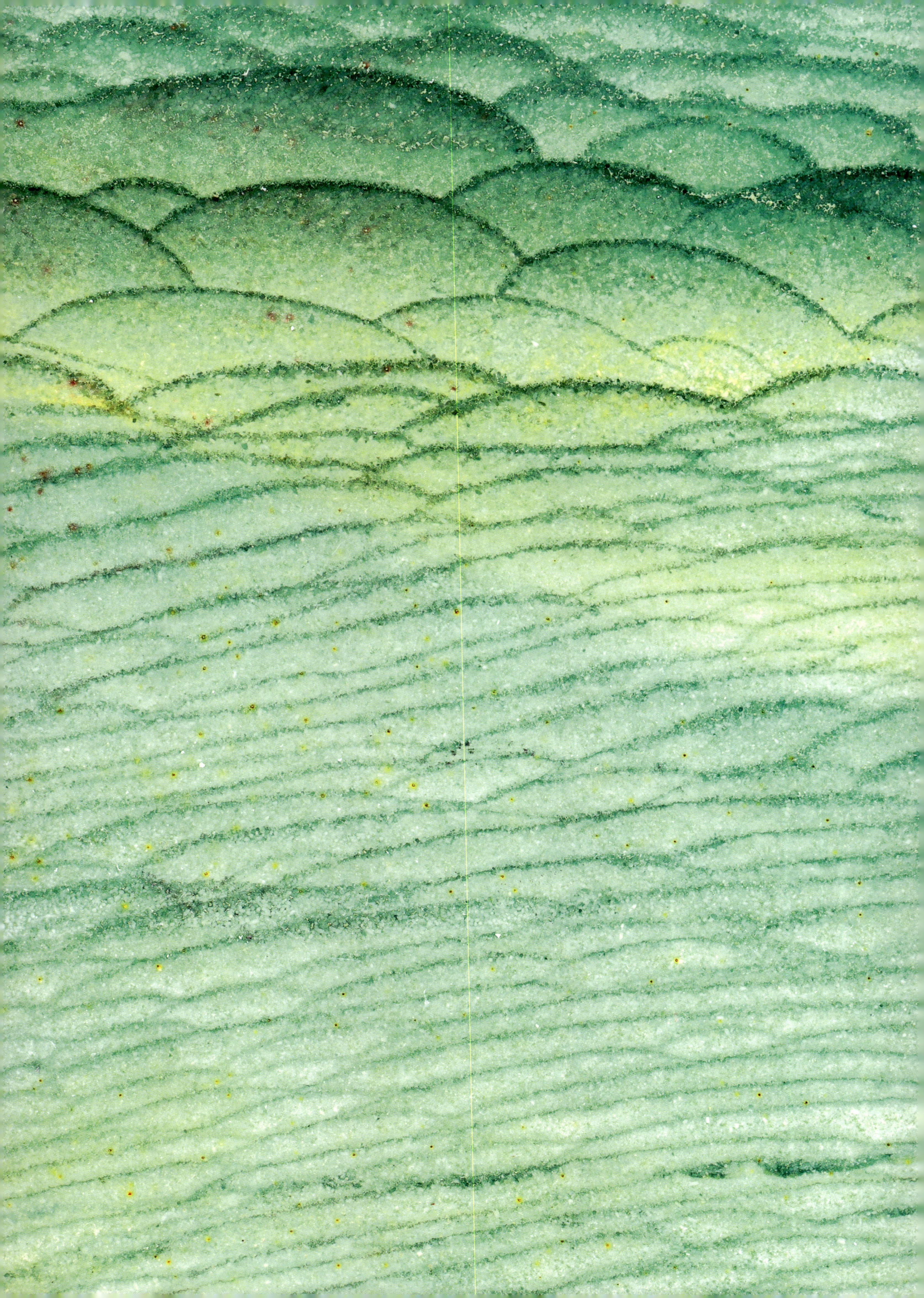

The examples we have considered so far have been no more than stars, rosettes, lines, and curves arranged in almost regular structures based on a pattern which though secret may probably be calculated. But there are other nodules free of all rhythm. No arithmetic may be divined in them. Large patches broaden into shining or glittering expanses dotted with shapes like tadpoles or little fishes, salamanders, alchemists' retorts with enormously long necks, strips of seaweed suddenly burgeoning into huge square bladders or volcanic bolts writhing upward and breathing eruption. Sometimes the markings take the form of dragons like those depicted on the silks of Asia, or spectral acrobats doing entrechats or attempting the splits. Priestesses in sparkling robes sheathing the mystery of their bodies take part in slow ceremonies and almost motionless processions.
Partly metallized deposits suggest underwater fishermen in harness and helmets brandishing the catch still transfixed on their harpoons; floating phantoms; tightrope-walking demons with eyes on stalks; a whole chilly sabbath of spirits, harpies, and bacilli. Narrow spindles with point downward, sometimes wide apart and sometimes joined together at the fullest of their curves, are divided at the top like the twoheaded birds on imperial banners, and part halfway down like mandragoras displaying their sexual organs. They seem both welded and extensible, like the folds of an accordion. You could easily think them about to expand, as if painted on some elastic material just going to be stretched. Long balloons full of grainy sap are joined together and fed through tubes.
The calcite crevices have sometimes been broken up by encountering knots or obstacles and having to find a new path. Fangs, horns, flakes of fire, sickles – all these may supplement or maim them, may either scarify or prolong. By a miracle these tortured effigies are entirely elegant. They produce scenes resembling tournaments, bullfights, and processions; also elemental columns of chromosomes, protozoa, and centipedes.
With the aid of chance or fancy, some of these images recall more precise and familiar forms. These make less demand on the imagination, echoing as they do images in the memory. One specimen undeniably presents a bull, its head turned awaiting the attack, penis tumescent, horns aggressive and eye sockets hollow, cheekbones suggesting the skull beneath the living skin. In another stone a catfish, with moustache bristling and a dorsal fin like a cleaver, has such a swollen head its body looks like a mere appendage. Instead of a tail it waves two diverging streamers, one of which turns absurdly, as if in a dream, into a kind of bird, which in turn is extended by a vast proboscis like that of a hawkmoth. Nearly all these animals and ghosts and stiff ungainly figures are merely conjectural. The same markings could just as well be identified, as in the onyxes, as the downstrokes and upstrokes of some elabourate calligraphy, or a subtle semblance of writing invented by some mad scribe, in love with shape but above meaning. And indeed the shapes themselves, whatever category they belong to, whether interpreted by the imagination or merely accepted by it as they are, whether they make use of symmetry, rhythm, or repetition, remain essentially forms, possessing what seems an inevitable harmony. Moreover they occupy the middle of the nodule, with its surface around them like a frame, just as a painter arranges his composition around the centre of the canvas which he is careful to separate quite visibly from the surrounding space.
The patterns in septaria consist of strictly plastic equipoises in which nothing is regular, and each of which is as unique and, in a manner of speaking, as personal as a deliberately executed work of art. While in the days of figurative painting connoisseurs used to frame landscape and ruin marbles as parallels to real pictures,

nowadays they should certainly choose septaria from among all the other natural forms to compare with many features of contemporary art.

Jaspers and Agates

Because they exist in such variety, agates present a great multitude of different images, though all of them are ambiguous and vague like patterns in the clouds. In order to fix them, the imagination has to play its part and cling to the image it has settled on. Moreover, the markings differ according to the kind of mineral they are found in. Similar motifs occur in agates from the same place, so that an experienced connoisseur may deduce the provenance of a specimen from the topography of veins, patches, pockets, and strata revealed by cut and polishing.

The image is often almost abstract and is at its most powerful when based on the elementary geometry of the circle. But – admittedly this caprice is a rare one – an agate will sometimes offer resemblances, suggestions: a hummingbird with an amethyst tail drinking from a flower as it hovers in motionless flight; tracks emerging on to the desert between the sheer sides of a mountain pass; lightning zigzagging across a stormy sky like a caliph's signature or a torturer's whip; green waves or scales or tiles, like the skin of a snake, like the roofs of markets and almshouses in Burgundy, or like the surface of the sea; or again, like a sea of clouds in a Japanese print, with rows of partly overlapping arches shadowing forth an infinity of white horses, the trembling of an aspen, an ebbing, flowing, breathing peace.

Another agate has an image of perfect simplicity: a single white shape on a background of sooty black, a rounded form suddenly rising to a point at the top, on the left. It looks like a little bird coming out of the egg, its body, with the wings still clamped to its side, not yet free of the shell but the young sightless head raised eagerly into the new dimension. As it struggles for its liberty it discovers space, transparency, an invitation to flight. It would have taken a painter of genius to invent the marvelously delicate edge of grey that shades the body just where it emerges from its prison, aspiring to the light and divining that it will be able to move through it. This immutable effigy in solid stone depicts a birth that is vain but fixed in a moment, a form and a colour that stir the emotions. The strong white shape is echoed in the wider curves, dark, almost imperceptible, in the body of the agate, as if the ambient air were disturbed by the sudden arrival of the new presence. Here is a creature quitting the stony oval which held it captive, crouched in upon itself and at once becoming keen, expectant, avid of the void. It knows it is about to burst forth and glide unaided through the air. And an almost colourless patch, defined by a few sober lines, has been enough to make us witness the miracle.

But likenesses remain farfetched. In most cases the markings to be found in agates provide only a rather unconvincing pretext for reverie. The imagination might just as easily refuse the bait altogether and decline to find a dubious solution to a false enigma. But experience soon shows it is not easy to maintain this aloofness.

However much I tell myself I am only looking at some irregular blue streaks crossing the amazed agate like the tracings made by a seismograph or a barometer gone mad, they still seem to splash up almost to the outer surface of the nodule with the transparency of mead or urine. However often I remind myself that the black undergrowth in the lower part of the cross section is only the usual ramification produced by dendrite of manganese, even while I am reducing things to their chemical constituents I cannot help descrying swathes of arctic light shining meagerly on inky lichens, a puny, struggling vegetation exhausted by rough winds and burned by frost.

Probably no images are utterly silent. Here a thin white line drawn by the finest possible point traces the outline of a perfectly round kidney of onyx: in its night of polished jade it imprisons, sparkling brighter than its own somber surface, the crystals of a frost instinct with darkness. Impossible to say what symbol haunts this primitive diagram, yet the sophisticated allegories from Florence, invented to express a complex relationship beyond words, are far less eloquent.

Onyx, with its dense deep black, is a special substance. It lacks the rich and glassy brilliance of obsidian, and I can well understand why it was not of obsidian but of onyx that an ancient lapidary said it reflected the shadows rather than the images of things. The grain of onyx is so fine that the veins it contains, utterly free though they are, can follow their whims with impeccable precision. In the best examples the lines are so regally sure of themselves that the need to seek for models is abolished. They express, they represent nothing but their own clarity. Without precedent, without meaning, they simply are – grandly adding to a world of appearances that does not contain their like.

One onyx may resemble, though, a votive stela like those that celebrate the fame of kings. Its shape, emphasized by a double border, is that of an irregular semicircle, set in an ample casket of quartz spines. Standing out against the shimmering dark of the stone, and following the shape of the containing nodule, there are marks like embroidery, suggesting ornamental writing. There are two breaks in the inscription, as if spacing out a brief dedication. The writing is so ornate as to be unrecognizable, like the Arab or Persian calligraphy traced on tiles to proclaim the greatness of God over the gates into the high mosques of Samarkand and Ispahan. But here the script is fluid, inscrutable, furtive, smoothing out the characters instead of sharpening their features: instead of spreading them out in bristling symmetries, panoplies of swords or fans of scimitars, twining them into coiling and uncoiling reptiles.

Suddenly you wonder whether this might not really be writing instead of images of a thousand other things: decorated snakes with gaping jaws as in Aztec fables; pale caterpillars swollen with lymph and latex and eaten up with gangrene; the sinuous, mazelike seams of the skull or in the whorls of the larger ammonites; or, closest comparison of all, the network of thin spikes which makes the king crab' s carapace both light and strong. To decipher such writing, if writing it is, would not mean trying to unravel an inextricable mass of lines and loops, but rather endeavouring to interpret anew some oft-repeated signs so turned in upon themselves that they refer only to their own form. In the middle is a larger, plainer motif, perhaps the name to whose hidden splendour the surrounding flourishes, titles, and complimentary epithets bear witness. Between the inscriptions the dark surface of the stone is covered with a tiny intricate pattern of meandering lines, which fill an unfathomable mineral grief with a mysterious shifting life.

But it is not an alphabet: it is a pattern without a message, like the wormholes made by insects in dead wood. One example is the almost closed chink formed by a pair of thin lips, insipid yellowish white against an intensely dark background, like the pale, anemic, reluctant vagina said to have opened in the womb of the Abyss at the beginning of time.

The jaspers of Oregon are probably unrivaled for the almost morbid complexity of their curved designs and their range of contrasting or merging colours: a graphic madness attained by no other mineral. Each one, however small, looks like a coloured lithograph, crammed and tumultuous as a picture by a schizophrenic. There is a continual movement to and fro between pattern and background, background and pattern, though the two are difficult to separate for the shapes in both are similar, the same purplish red predominates in both, and not even the smallest space is left unoccupied. It is the ultimate pin's head, suitably caparisoned.

We have here a universe of scrolls, branches, pleura; from them flayed countenances emerge, muscles laid open in their cavities of bone. There are lopped-off breasts, the mutilation twisting the raspberry nipples aside; there are the bodies of frogs, crucified by the galvanic current, their limbs splayed out by the shock, their skin turned blue and flabby by the violence of the spasm. Elsewhere we find a loose array of small tools, toys, and useful objects: bobbins, spools, shuttles, tops, drawer knobs. In the distance, but still quite near, are dunes, stretches of sand rhythmically modulated by the wind, a screen of hills, a host of weathered peaks with the geological strata bared, fleecy rowers as still as tropical clouds. The stone may be purplish blue, lilac coloured; yellow turning green; the complete range of a bruise. It may be like a swollen sea of thick, almost solid bubbles, resembling an upsurge of sinister seaweed or an eruption of boils or buboes on an infected skin.

Then again, as if by the adjustment of a viewfinder, the scattered elements let themselves be identified: we see an eye devoid of lid or lashes, or an empty socket with the freshly removed orb dangling like a wet rag or an oyster torn from its shell; gnarled and ringed phalluses, swollen and purple, without their foreskins, the glans all wrinkled; rotting shelless mollusks, the colour of lichen or of gobs of spit; kneecaps and knuckle bones softened by acid to a cloudy, wobbling jelly; intestinal worms glistening with the biles and juices digesting them; a jumble of passages, rumbling innards, excited vulvas, striated tendons; pale partial globes jointed like the knees and elbows and hips of celluloid dolls.

A mauve and lunatic life, proliferating without law or limit, feverishly breeding tumors and goiters; a ravenous, shifting universe in which details are so clear it is almost endless. Wounded flesh shows how this monstrous realm works, idly limned by imperturbable stone which neither feels nor knows.

Still, it is a charming image, full of invention and surprises, for the easy-going connoisseur who confines himself to colour and composition.

At the other extreme the wonder stones of Idaho, a different kind of jasper, are remarkable for their sobriety. In these the pattern often consists of concentric ellipses swelling out on either side of a vertical axis. Where these ellipses seem to overlay one another they are paler, as if superimposition made them fade. The wider they spread, the darker ochre they become, until the outer edges are almost brick red. There are streaks in them which appear even deeper in colour than the reddish brown of the background. They proclaim the circuits of planets or electrons around their invisible centres or nuclei; they are an image of the fundamental yet simple law of gravity that links physical bodies together at every level of the universe. They are like the rings of armillary spheres, with their zodiacs and ecliptics and equinoctial zones – bracelets for cosmographers or nuclear physicists. They reflect the phantom revolutions which, alike on a vast and on a microscopic scale, unflaggingly repeat the same pattern. This is the blueprint of nature itself, both hidden and revealed in a nodule of silica, making known the blazon of the universe, the constant figure which governs it in its entirety. But in order to be moved by the pattern ironically inscribed thus in the heart of a stone, one needs to know already the secret it unveils or recalls, one needs to have learned from scientific works of the thousands of patterns which this one brings together, and without which it would remain what it really is – chance curves providentially assembled by another chance and randomly coloured by metallic deposits.

The vision the eye records is always impoverished and uncertain. Imagination fills it out with the treasures of memory and knowledge, with all that is put at its disposal by experience, culture, and history, not to mention what the imagination itself may if necessary invent or dream. So the imagination is never at a loss when it comes to making something rich and compelling out of a subject that might almost seem an absence of all life and significance.

Just as I am about to leave this world of signs, of pretenses, of solicitations either subtle or urgent, one last agate with complex and unusual markings leads me yet further along the road of reverie. The translucent interior of the specimen is enclosed in a shell that is rough and opaque, with the pale beige, hazel, and walnut colouring of silicified wood. Slightly to one side of the middle of the cross section is a window of quartz needles: these spines are mauve toward the tip, and form around a central hole a sort of diamond, three of whose sides are convex while the last is concave. The scintillating prisms of the quartz are in turn surrounded by a wide band of azure chalcedony, itself edged by a border of red or orange. The diamond seems to emerge somewhat battered: one of the corners is drawn out to a peak, another twisted into a kind of pouch. Around it on all four sides lie a series of closed quadrilaterals outlined thinly in cherry red and royal blue alternately. They never touch the points of the central diamond but proliferate outward briefly and then die out, their lower extremities prolonged by luminous strokes that are swallowed! up in the warm confusion and amber smoothness of the agate. They are like rough and broken reflections, among which, it is impossible to say which is the original and which the derived image. They simultaneously balance and differ from each other, like the deliberate irregularities in an oriental carpet. The lines that bound them suddenly stop, while the bands in the stone, which they now no longer enclose, go on echoing and relaying the shape of the main diamond until they reach the stone's ligneous periphery. They gradually blur the shape of the central lozenge, but throughout it still governs the quivering symmetry of the whole composition.

It is impossible not to admire the riot of colours, the magnificence of the material itself. But these are not exceptional in agate. What is surprising is the tortured symmetry, the broken system of angles and polygons striving to achieve an overall pattern: a geometry without the strength to finish what it has begun, an attempt at organization which finds fulfilment only in the structure of the septaria. The image which results from the combination of those shimmering colours and which ambition, pathetically striving against a substance that thwarts and engulfs it, induces in the beholder the same surprise, almost warning, as that produced by a sea louse, a water scorpion, an anteater, or any other creature left behind at some crossroads of evolution.

One day some bold being invents – or has invented in him – a form which may be valid at the time but which is soon set aside in favour of a simpler and more elegant solution. The brave discovery survives through inexplicable negligence on the part of those stern powers that usually eliminate the fantasies of a moment, even though that moment may last thousands of years. Such forms endure only to bear witness to life's mistakes, to remind nature of its monsters, its botched jobs, its blind alleys.

The strange markings in an agate lead me suddenly into wild extrapolations, seeking, in the obscure workings taking place within a stone at the dawn of time, traces of similar abortions. By their very strangeness the failures they perpetuate become for me so many speaking portents, or at least emblems. They somehow announce the coming, in the distant future, of a species that makes mistakes, a being in whom freedom and imagination, together with their necessary disappointments, will be more important than successes due to infallible and inevitable mechanics. They presage new powers, imperfect but creative.

Such aberrations exert a special fascination on man. They seem to be manifestations par excellence of what I have ventured to call *natural fantasy*. Perhaps their undeniable, unfailing, and yet mysterious attraction lies in the fact that, through some dim reversal, they assert the right a doomed nature has won to the gratitude of its latest, grudging heir.

Man has unknowingly inherited a capital made up of immemorial audacities, unsuccessful risks, and ruinous wagers, an endeavour which though for long persisted in vain was one day to foster in him a new, rebellious grace, combining hesitation, calculation, choice, patience, tenacity, and challenge. I can conceive of some divinity, some total intelligence that is panoramic in the widest sense of the word, capable of contemplating in one purview this infinity of vicissitudes and their inextricably complex interactions. Such hypothetical cosmic consciousness would not be surprised at the existence of a lasting and inalienable collusion between this series of fertile abortions and their ultimate beneficiary. It would seem to it inevitable that a secret affinity should allow the heir to recognize, among the daunting mass of nature's ventures, those which, though they did not succeed, opened up for him, through their very failure, a glorious way ahead.

Limestones from Tuscany

The whole surface of this specimen of graphic limestone is crisscrossed with long, tapering triangles and quadrilaterals of extraordinary clarity. Of every shade from buff to brick, they suggest great polygonal grasshoppers packed close together and intermingled, wing cases whirring and long legs splayed, the head of one clinging to the abdomen of another. The voracious swarm of tangled insects looks like a wallpaper design, recalling those carved ivory balls from Japan, covered with a mass of rats or crabs devouring one another in a perfect, spherical and! horrid continuity. But here all is flat, angular, and diagonal. Running across the welded heap of bodies, sometimes separating them with a bold stroke, then suddenly piercing them right through, is a web of thin lines like nerves or stiffened arteries, the thinnest metallized, the others made up of tiny crystals. They are mat and dull so long as they do not reflect light, but as soon as the stone is placed so that it catches a beam, the network of lines lights up. Among those crickets, packed like sardines, speeds hair-fine electricity. A cat-o'-nine-times-nine-tails laps them in its supple lashes, its darts of mercury. The whole surface flashes. So, in mountain country when the snow melts, meadows are streaked with springs rushing down from the shadowy hollows where the neve forms. It is a fiesta of droplets and foam, a wild rush of silver in a panic rush downward, leaping, wearing itself out, soaked up slowly at last by an already saturated earth. But on the polished surface of the stone no exhaustion lies in wait for this multitude of fiery rills. One gesture sends them to sleep, another reawakens them: the spring sparkles and plashes anew, an incandescent lava fretting its ardent way through tenuous furrows yielded by the stone's fine resistance. Above this tissue of light, in a small space sheltered from the oblique rain, is a distant disk, a minute spot of dull leaden hue imaging a melancholy Saturn.

The surface of the stone is not always striped in this way with violent diagonals like stinging showers driven by hostile gusts, and weirdly polarizing the light. Other types of Tuscan limestone, also related to ruin marbles, sometimes depict what look like elabourate landscapes. They show misty forests of tall pines like those in the paintings of Böcklin, but half obscured by the vapors of dawn, dimmed by a magical, ghostly haze. Others show wild and withered clearings invaded by scrub, with short tufts of scrawny grass growing up between the pebbles. Above these stubborn thickets a desolate sky is flecked with slanting black hail. Other examples have only languid patches slowly spreading and changing colour as they do so, like tars or mineral oils of different densities. Sometimes stones with triangular markings or patterns resembling vegetation present extreme examples of the tricks minerals play on the imagination. I shall try to describe an example of each of these two kinds, in which the image is a kind of mirage. The illusion – in the one case, of a castle; in the other, of a portrait – is so precise, detailed, and compelling, yet at the same time so improbable, that the beholder feels drawn to try to analyze the mechanism that produced it.

The Castle. The outline of a huge, shining brown castle stands out against a pale sepia background. This background becomes mat in an oblique light, while the building shimmers with an almost metallic brightness. The values change, but the shapes remain the same. Deep ramparts link the different parts of the edifice, in the middle of which a stepped tower dominates the rest. What we see is a cross section, without thickness or perspective, a mere elevation of the imaginary building. Whatever height one attributes to it, it is still looked down upon, overshadowed, by the broad, drooping leaves of great tree ferns, unfurling their lovely lace high above the towers. The beholder wonders what tropical vegetation can have grown such enormous foliage, making a palace look like a doll's house. The eye hesitates; not knowing which scale to choose, it alternately magnifies the fern and diminishes the building. Some birds are eddying about in the right-hand part of the sky; to the left there is only one bird, but a huge one, with wings outspread and neck pointed earthward, about to swoop down on the irregular parapets and the strange figures moving among them. For the castle is inhabited: on every parapet, in every moat and window, climbing up the walls, are parallel shapes all facing in the same direction and frozen in the same attitude. These clearly defined though clumsy figures, like the little men children draw, all stand in silhouette, facing to the right, stretching their arms out in front of them as if they were blind, reaching into space or toward the nearest wall. They too are only cutouts, as in a shadow show, and their lack of depth adds to the unreality of the scene. What are these flat creatures looking at? Where are they going? Is their attitude one of protection or of prayer? On the far right there is a kind of bridge, and beyond it the only figure different from the rest seems to be waiting for them.

This shape is not seen in profile, and a patch of white gives it the rudiments of a face. The whole scene is rent three times by heaven's bolt, as the crazy universe is struck through by a white zigzag of lightning. This stone looks just like a picture. Everyone to whom I have shown it has at first taken it for the work of some naive or unskilled painter, a child or a tyro. Only when you examine it closely do you perceive that it is a kind of natural picture. Some people even then find it hard to accept and think it must be a cheat, so inconceivable does it seem that chance alone could produce something so similar to the work of man. But in fact there is nothing, or almost nothing, prodigious in the elements that make up the image here. The "giant ferns" are merely dendrites (though admittedly in an unusual form) such as may be found in abundance in many minerals. The "birds" may be reduced to cruciform markings. The lines suggesting broken façades are not so precise and evocative as those recalling ruined cities in the Florentine marbles. Only the little men are really astonishing. Simultaneously alike and different, all of them stand solidly on both feet, head high, arms either exploring space or bearing some invisible offering. Their shape is well defined and can only suggest a human form. The strangest thing about them is perhaps that the silhouette is repeated, and that every time its feet are on the ground, while the body stands out entirely vertical in the middle of an empty rectangular space like a window or a door.

This is really the only truly amazing coincidence here, and even it is not inexplicable: for it is never impossible, it is even inevitable, that one form should resemble another. Admittedly in this case the form – the human body – gets special treatment: it appears not once but ten times, always in the

same attitude and always enhanced by a frame. It is this that gives rise to doubt, to the persistent idea that only deliberate design could explain such a pattern. We have now pinpointed the oddity around which the rest of the interpretation has organized itself: the ferns, the palace, the birds, and the lightning, this last just an ordinary veining that might be found in any piece of marble.
Here is neither miracle nor mystery, just an extraordinary combination of signs which have no meaning but which are swiftly given a meaning that the ensnared imagination finds it hard to withhold.

The Portrait. Diverging lines fragment the surface of the stone without impairing its continuity, breaking it up into sections that juxtapose different shades of the same colour, usually beige, green, or black. As the lines emanate from several different points and are contained in a very small space, they soon intersect, making an irregular web of triangles, trapeziums, and diamonds, some slightly darker and some slightly paler. It is as if light were an artist, subtly yet capriciously trying out the range of its own intensities.
Nature usually contents itself with a chaos of lines and angles overlapped and overlapping. Their thrust, at once parallel and athwart, suggests the idea of sea waves or mountain peaks, as represented in simplified form by a geometrician trying to understand the laws governing their rhythm – a kind of diagram, as of a man walking, a horse galloping, or a *Nu descendant l'escalier.* Sometimes the lines are scarcely visible. Confused patches of green lining valley slopes depict mistshrouded forests beneath a grey sky. As the structures are always rectilinear, a rarer chance may bring forth human faces, like those portraits in which the painter has tried to disassemble his model's features in order to reconstruct resemblance in terms of straight lines, salient and reflex angles, and simple polygons. It is an unlikely phenomenon, but it does occur. It only remains to see how.
I have before me, captured in stone, a face that attempts the exploit described above. It stands out in three-quarter profile, grey against an ochre background. The slab is thin, yet on the other side the image has already started to grow dim and indistinguishable. But on this side the face, beneath its hood or felt hat, is crude, sharp, rough hewn as if by some rustic image maker. The semblance of a portrait is the result of a complex cross-hatching of the available space. Some unbroken lines cut right across the width, most of them veering slightly to the right. The highest one divides the lower part of the headgear from the shadow it casts over the eyes. The lower ones enclose the bust, between the shoulders and the waist, where the grey pigmentation stops. Between the two zones thus delimited, diagonals outline the steep planes of a haughty visage. The only areas pigmented with grey are those lying between on the one hand the almost parallel lines marking out the different levels of the portrait area and, on the other, the slanting lines coming up from the base and cutting across the others like the spokes of a fan.
A third set of lines come from the four sides of the stone, thrusting strange beveled tracings right into the central grey area. It is this finer, encroaching network which, by inserting its spears into the ashcoloured part, suggests the contours of a face. A bright yellow streak thus separates the square chin from the line of the shoulders, stiff under starch or armour. Another of these secondary lines emphasizes the prominence of the nose; another darkens the eyebrow and socket above the only visible eye; a fourth indicates the near right angle between the slope of the cheek and the terrible lower jaw, the quintessence of obstinacy and disdain. Others, less clear, suggest the meagre white shape of bloodless lips. Together these lesser lines mark out the lights and shades that give the dust-coloured countenance modeling and life.
Here again, no miracle. The portrait is the result of a composition formed by several series of straight lines variously oriented. Their random intersections form cells of different shapes and sizes.

A second chance complements the first, and the central boxes are coloured grey by metallic salts suspended in the limestone and compartmented by delicate seams in the stone. Hence the contrast between the colours on either side of the bolder lines defining the main motif. The third and by no means the least chance, infinitely improbable though much less so than that of the monkeys typing out the *Odyssey*, is that which makes the image resemble a person in some sort of hat and ample cloak. The shape is impressive, of course, but hardly more so than those one suddenly identifies in the clouds, where the eye recognizes one after the other giants, animals, and battles. So far, then, nothing out of the ordinary. But at this point two new factors intervene. First, in the clouds and elsewhere the likenesses, as I have said, are elusive. The incalculable chance that produces them is never more than a fast or slow succession of unstable states, shifting shapes which fade away as soon as seen. The beholder who catches them on the wing knows they are going to escape. The appearance he was beginning to identify changes, dissolves, and in the time it takes to recognize one image a second and a third, as vague as the first, are born and in their turn disappear. It was all a dream. The insubstantial marvel was impossible to grasp; it was not a marvel at all. Its brief manifestation was too fleeting for it to make an impression on the memory. It lacked duration, presence, existence even. Is it certain it was ever there at all? It remains a dream or an illusion, like a fleeting reflection on watered silk. Wonder can only arise and increase when a spectacle endures, survives perception, and is ultimately seen to be less transient than the ephemeral being who came upon it.

But in a stone the image – every image – is fixed, as if the thickness of the mineral preserved the cloud, the flame, or the waterfall at every moment of its kaleidoscopic metamorphosis. Each image is an immortal witness, recorded for a long period of time: forever, measured against the brief human season. So all one has to do is get hold of an outstanding example, by trial and error or the favour of fate. The flowing currents have suddenly become motionless, all of them paralyzed, indestructible, though within reach of the cutting diamond or the polishing wheel. Now the stone has only to be cut and buffed, though this must be done along the best axis and just to the point where the marking yields its utmost.

A careful Florentine craftsman divided the stone I am now thinking of – the portrait stone – into thin layers so as to be able to follow the changes in the markings. Just once the saw went through in the right place. Just once it revealed on either side the phantom portrait which had been hidden for thousands of years and which is only a portrait, or even a sign, if the beholder collabourates. Before and after it is but an indistinct image in the process of becoming another image, another portrait or perhaps, as in the previous specimen, a hallucination of palaces, ferns, birds, and Lilliputians.

What we are dealing with is an illusion not merely of a human face but of a picture, that is, of a human creation depicting a human face. The impression that what we are looking at is a picture derives from the presence of a kind of frame, perfectly represented by an almost regular fringe of dendrites. Although they are uneven and interrupted their fine sprays nevertheless border the unyielding shape of the rectangular stone as if they were meant to enclose and enhance a real painting. A further refinement is that almost all the way round a thin line separates them from the background of what one is tempted to call the work itself.

This is no supreme coincidence or revelation. Between the angular effigy and a human face as perceived by the eye there is a distance: the so-called portrait is composed entirely of intersecting straight lines. But this only reinforces the analogy with works by such painters as Villon, Picasso, Braque, Juan Gris, Marcoussis and many others who, at one point at least in their careers, set themselves the task of dismantling the objects they perceived, and especially faces, and reconstructing them in terms of volumes and elementary planes. Just as, during a truce, negotiators meet in no-man's-land, so here a network of rigid lines comes together after a thousand anonymous and indifferent hazards to depict a face, echoing a structure deliberately invented by an artist so as to rebuild a living countenance in terms of abstract laws. This sort of coincidence is not an illusion; it is a warning, a signal. It bears witness to the fact that the tissue of the universe is continuous, and that in the vast labyrinth of the world there is no point where apparently incompatible paths, from antipodes much farther apart than those of geography, may not intersect in some common stela, bearing the same symbols and commemorating unfathomable yet complementary pieties.

The face is merely an accident. Yet another kind of sign would not have been so eloquent. I meditate upon its furtive, shimmering lights and shades; on the interplay of parallels and obliques; on the geometry first set forth and discovered by an ancient caprice. A chance created it, this unpredictable precursor of what would one day be the fragile human face, itself not at all indispensable to the world's variety: by another chance, it too might never have emerged from the infinite reservoir of forms. Who knows whether this tumult of triangles inscribed in stone, first brought about by nature and then invented by art, does not contain one of the secret cyphers of the universe?

At such moments it seems to me I see why these images exert such a powerful fascination over the mind; I seem to detect the underlying reasons for the unwearying and irrational zeal that makes man give a meaning to all appearances devoid of it, to look for parallels everywhere, and to create them where they do not already exist. I see the origin of the irresistible attraction of metaphor and analogy, the explanation of our strange and permanent need to find similarities in things. I can scarcely refrain from suspecting some ancient, diffused magnetism; a call from the centre of things; a dim, almost lost memory, or perhaps a presentiment, pointless in so puny a being, of a universal syntax.

Enter Life: The Other Writing

Life appears: a complex dampness, destined to an intricate future and charged with secret virtues, capable of challenge and creation. A kind of precarious slime, of surface mildew, in which a ferment is already working.

A turbulent, spasmodic sap, a presage and expectation of a new way of being, breaking with mineral perpetuity and boldly exchanging it for the doubtful privilege of being able to tremble, decay, and multiply.

Obscure distillations generate juices, salivas, yeasts. Like mists or dews, brief yet patient jellies come forth momentarily and with difficulty from a substance lately imperturbable: they are evanescent pharmacies, doomed victims of the elements, about to melt or dry up, leaving behind only a savour or a stain.

It is the birth of all flesh irrigated by a liquid, like the white salve that swells the mistletoe berry; like the semisolid in the chrysalis, halfway between larva and insect, a blurred gelatin which can only quiver until there awakens in it a wish for a definite form and an individual function. Soon after comes the first domestication of minerals, the few ounces of limestone or silica needed by an undecided and threatened substance in order to build itself protection or support: on the outside, shells and carapaces, and on the inside, vertebrae that are immediately articulated, adapted, and finished down to the last detail. The minerals have changed their employ, been drawn from their torpor, been adapted to and secreted by life, and so afflicted with the curse of growth – only for a brief spell, it is true. The unstable gift of sentience is always moving from place to place.

An obstinate alchemy, making use of immutable models, untiringly prepares for an ever-new flesh another refuge or support. Every abandoned shelter, every porous structure combines to form, through the centuries and the centuries of centuries, a slow rain of sterile seeds. They settle down, one stratum upon another, into a mud composed almost entirely of themselves, a mud that hardens and becomes stone again. They are restored to the immutability they once renounced. Now, even though their shape may still occasionally be recognized in the cement where they are embedded, that shape is no more than a cipher, a sign denoting the transient passage of a species. Unceasingly the microscopic roses of diatoms, the minute lattices of radiolaria, the ringed cups of corals like tiny bony disks with countless thin spikes resembling circles of converging swords, the parallel channels of palms, the stars of sea urchins – all sow seeds in the depths of the rock: the seeds of symbols for a heraldry before the age of blazons.

Meanwhile the tree of life goes on putting out branches. A multitude of new inscriptions is added to the writing in stones. Images of fishes swim among dendrites of manganese as though among clumps of moss. A sea lily sways on its stem in the heart of a piece: of slate. A phantom shrimp can no longer feel the a.ir with its broken antennae. The scrolls and laces of ferns are imprinted in coal. Ammonites of all sizes, from a lentil to a millwheel, flaunt their cosmic spirals everywhere. A fossil trunk, turned jasper and opal like a frozen fire, clothes itself in scarlet, purple, and violet. Dinosaurs' bones change their petit-point tapestries into ivory, gleaming pink or blue like sugared almonds.

Every space is filled, every interstice occupied. Even metal has insinuated itself into the cells and channels from which life has long since disappeared. Compact and insensible matter has replaced the other kind in its last refuge, taking over its exact shapes, running in its finest channels, so that the first image is set down forever in the great album of the ages. The writer has disappeared, but each flourish – evidence of a different miracle – remains, an immortal signature.

Nota bene

As this book has turned out to be a series of reflections on what I have learned from familiarity with certain stones, I have decided after some hesitation to describe and reproduce in it only those I have known well, those I have often looked at, handled, and caressed.
So the specimens that figure here all belong to my own collection, with the exception of certain seventeenth-century picture stones photographed in the Opificio delle Pierre Dure in Florence [not shown here, *publisher's note*].
In the interests of clarity I have employed the words agate, chalcedony, and onyx in their current sense and not in that which they are given in mineralogy, which is not entirely consistent, anyway. Thus agate is the term generally used here for the various types of microcrystalline silica; chalcedony denotes the milky or blue grey variety, often with fine parallel stripes; while onyx is reserved for the black variety with white veins, never the kind of greenish or brownish calcite commonly so-called in the trade.Jaspers are also silicas, but without well-defined demarcations and without the translucency of agate.
Last, I fear I have sometimes used the term septan'a for nodules which have only the appearance of such phenomena but which do not really contain fissures due to contraction.

Notes

1. *M. I. Galfarel,* Curiosités inouïes sur la Sculpture talismanique des Persans: Horoscopes des Patriarclres et Lectures des Etoiles *(Curious Facts about the Talismanic Sculptures of the Persians: Horoscopes of the Patriarchs and Readings of the Stars), (Paris, 1629).*

2. Recueil des Monuments, *1: 116.*

3. De marmore variorum locorum in specie Florentino, *quoted ibid.: "Hae ruinae sum lusus nacurae ex coalicu fluorum terreorum variorum colorum, qui olim liquidi et molles fuerum, demum vero, successu temporis, accedeme spiricu, terram subtilissimam coaguleme et gorgonico, petrificati et in marmoream soliditatem mutati?"*

4. *Ibid, p. 117.*

5. *E. M. L. Parrin,* Histoire Naturelle des Mineraux, *Year IX, 3:280-82; plate p. 280.*

6. *Paris, 1957, pp. 47-72; figs. 30, 31, 35, 39.*

7. *Quoted in Baltrusaïtis, p. 56.*

8. *Blaise Pascal,* Pensées, *frag. 134, Brunschwig.*

9. Cahiers du Musée de Poche, *no. 1, March 1959; Ans, no. 788, July 21- 27, 1960;* Meduse et Cie, *Paris, 1960, plate 2.*

10. *Lin-Yu Tang,* Chinese Childhoods, *(Paris, 1960) 2: 120.*

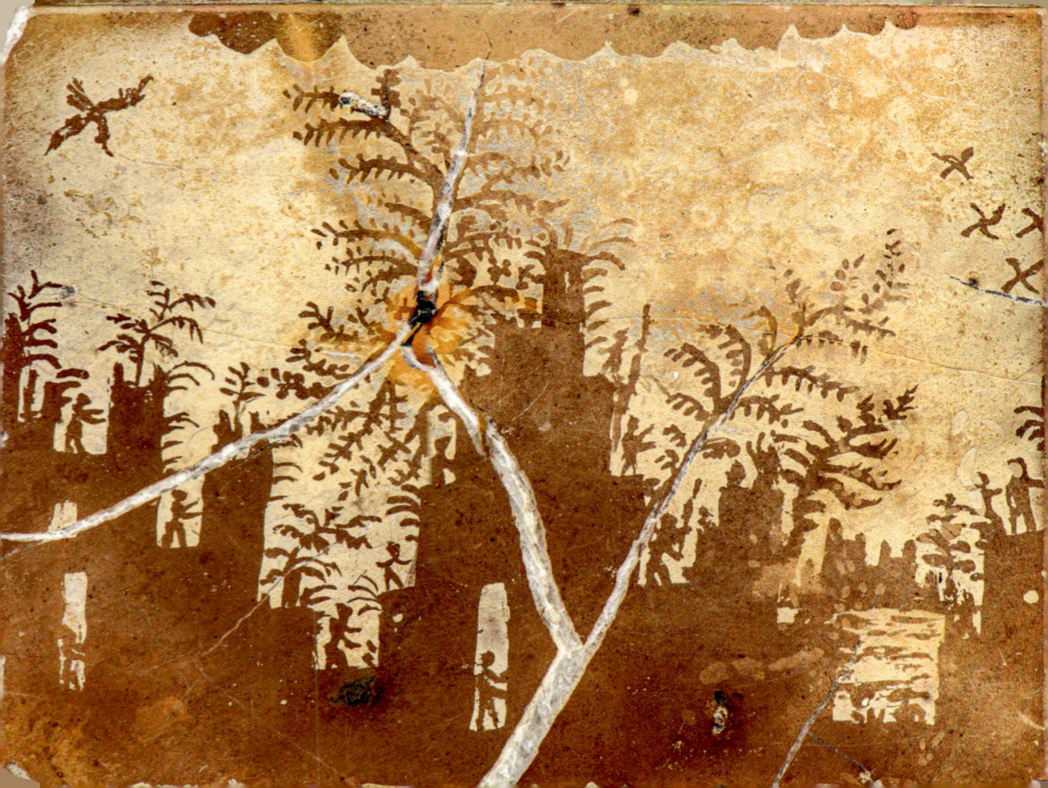

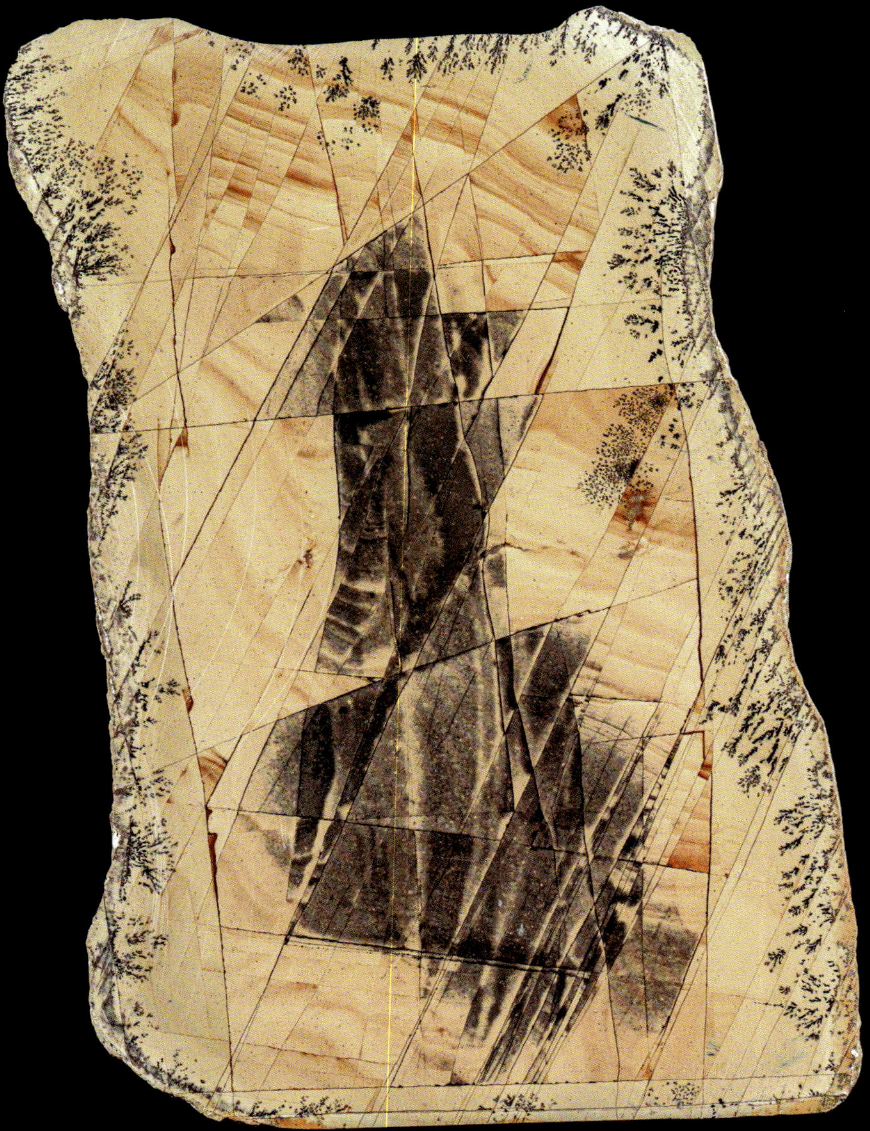

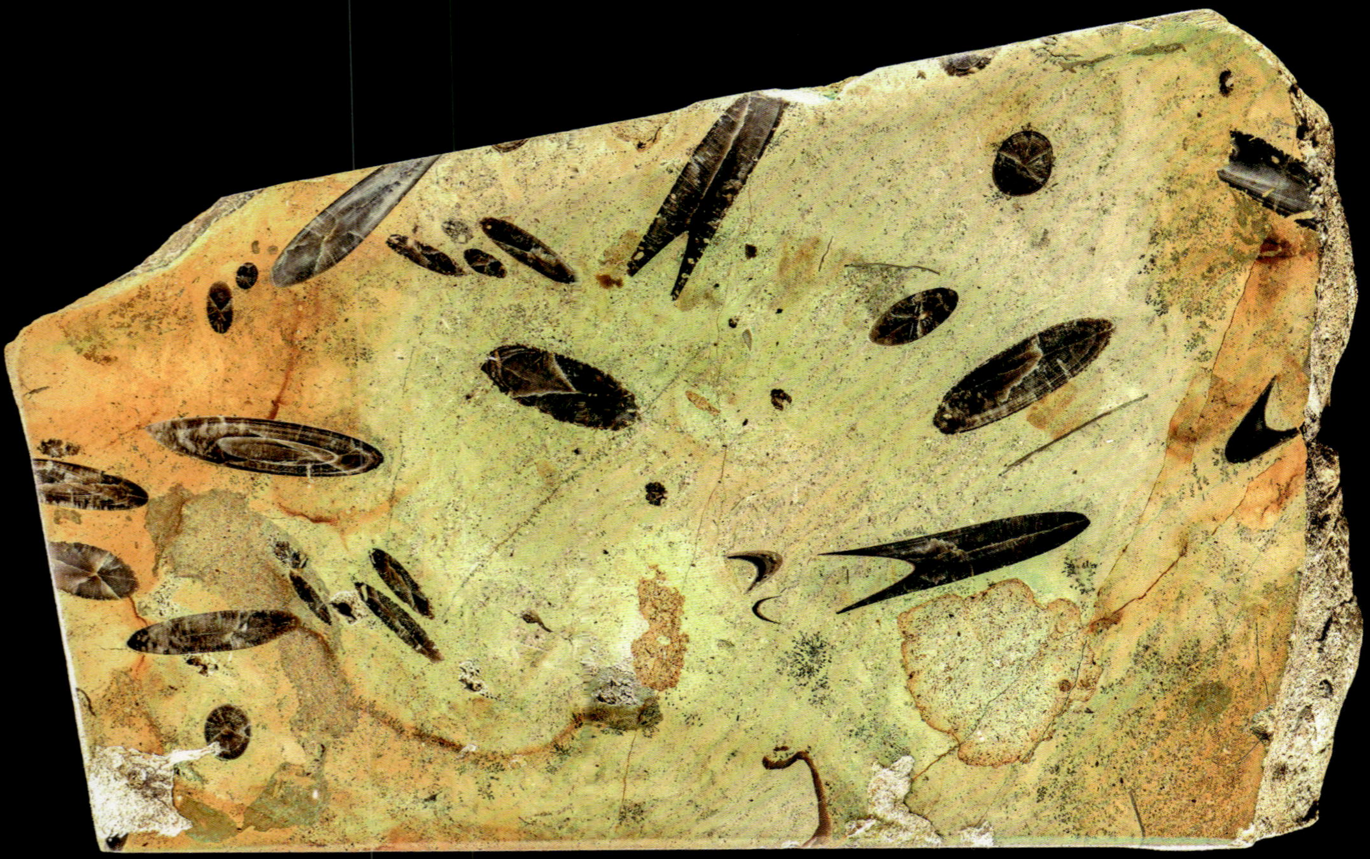

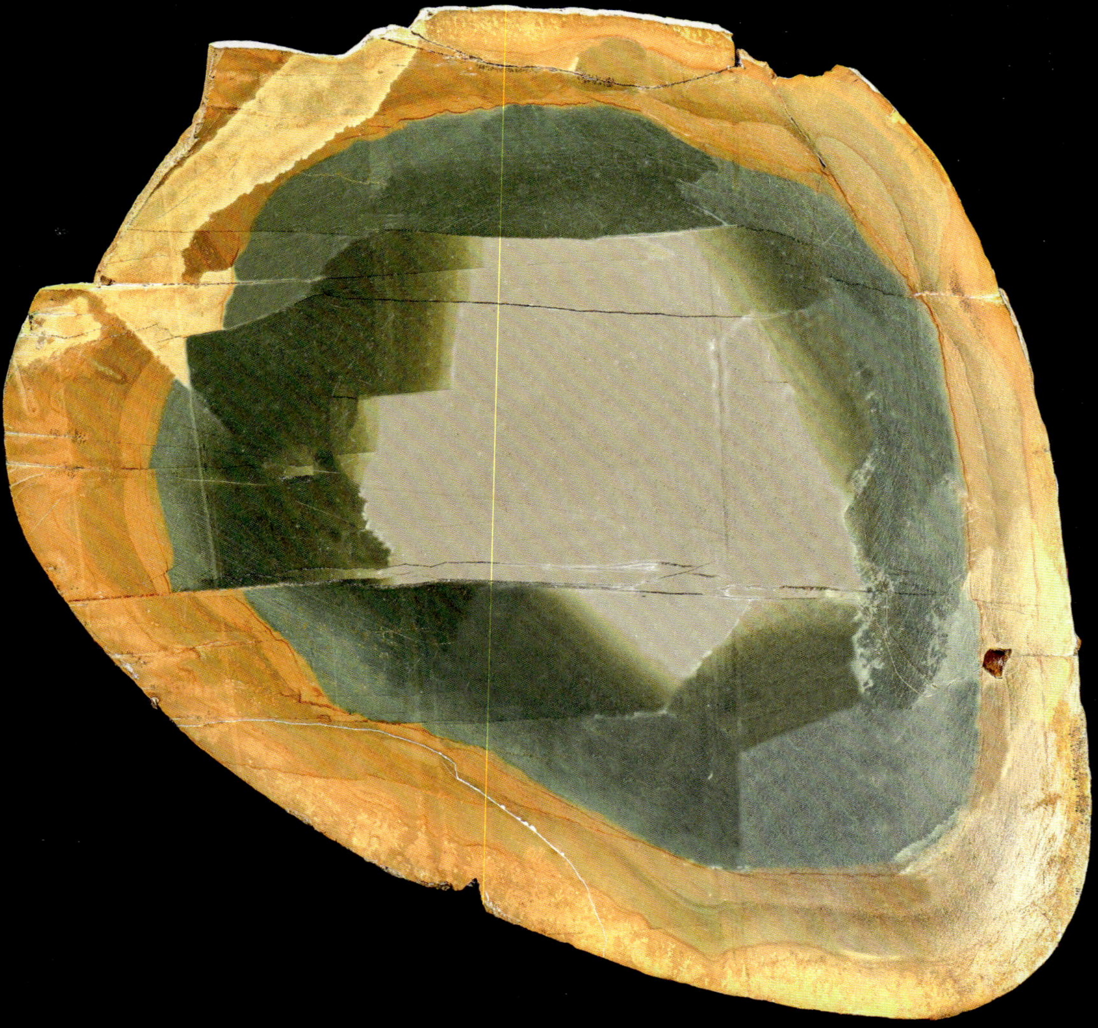

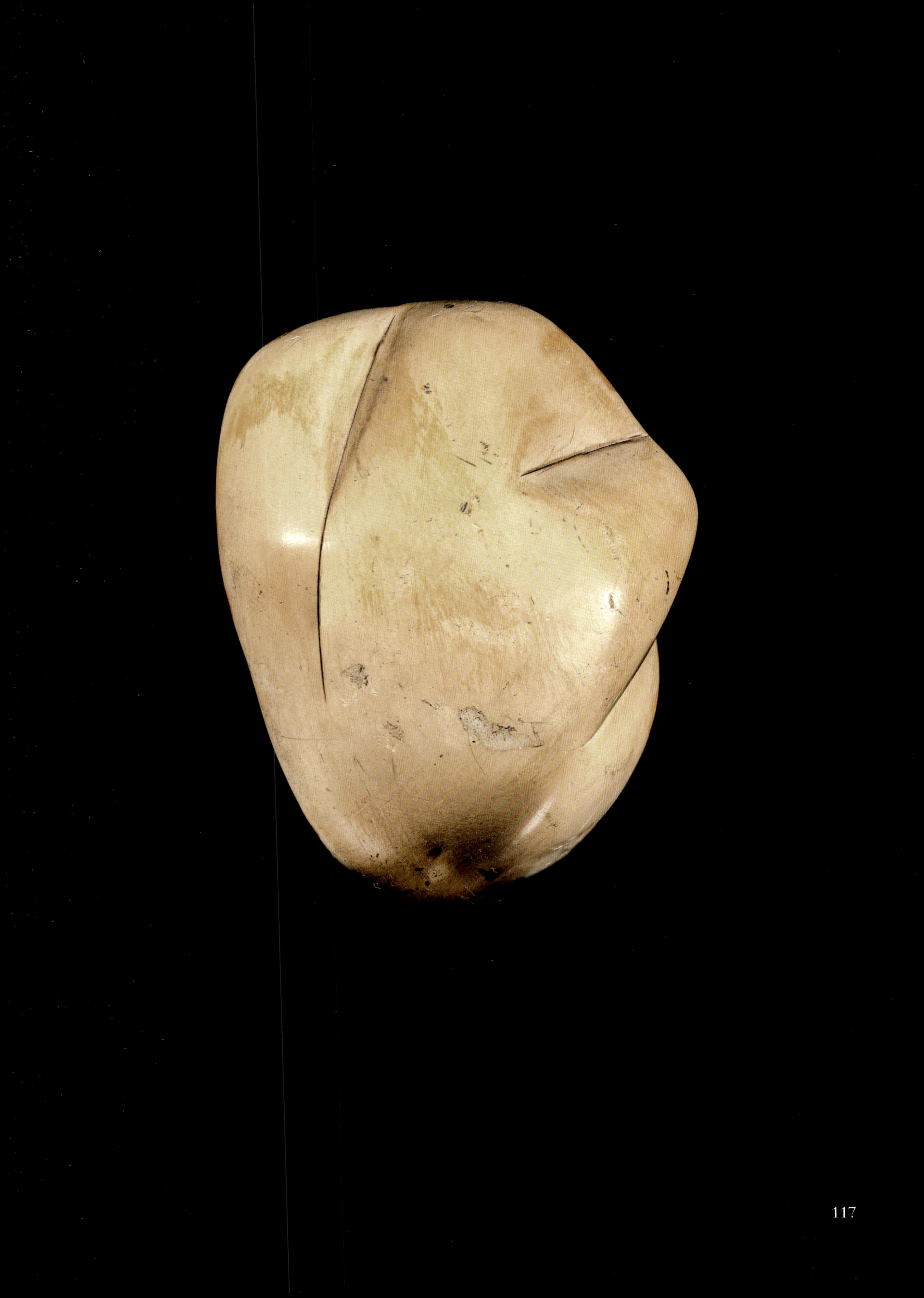

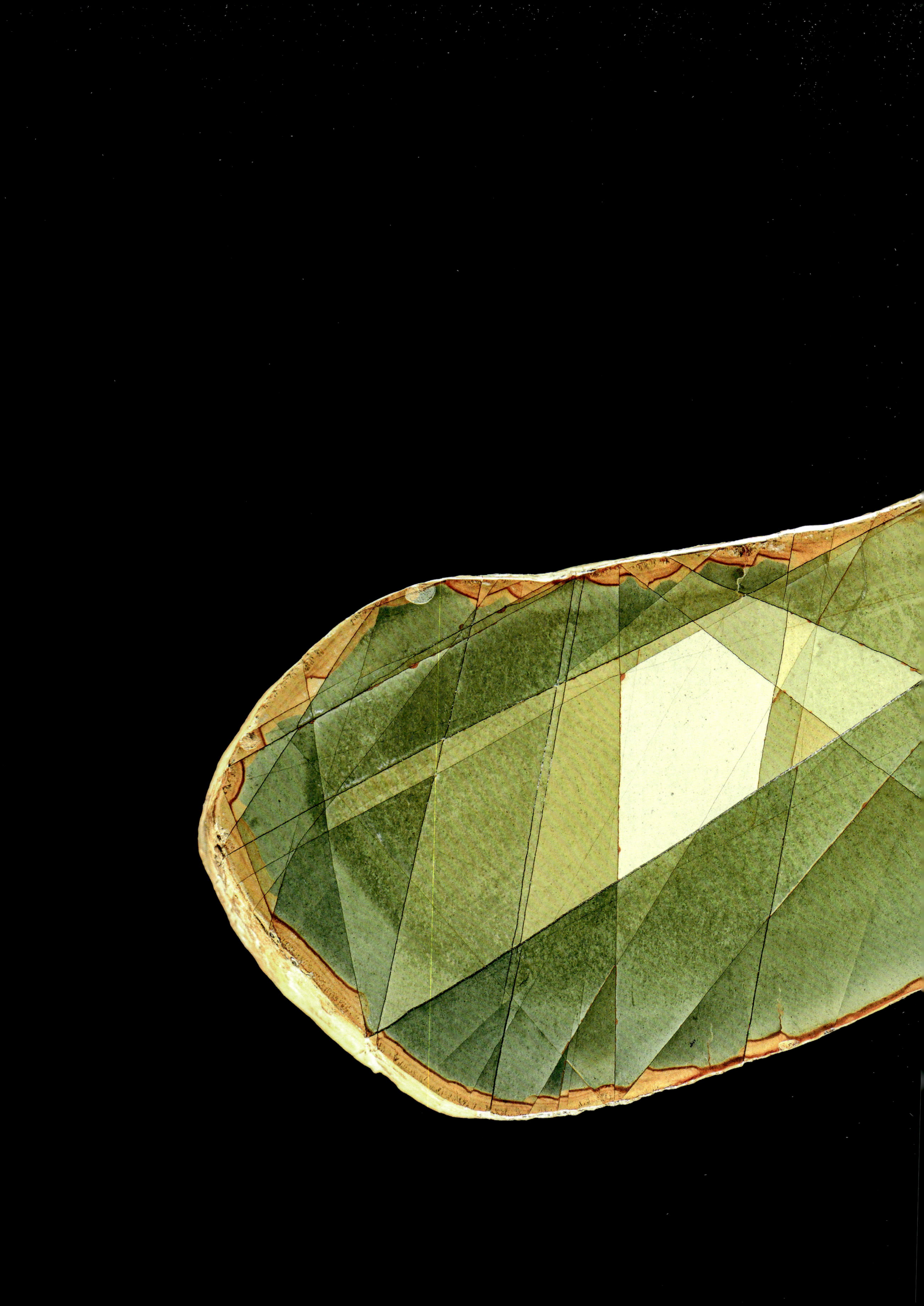

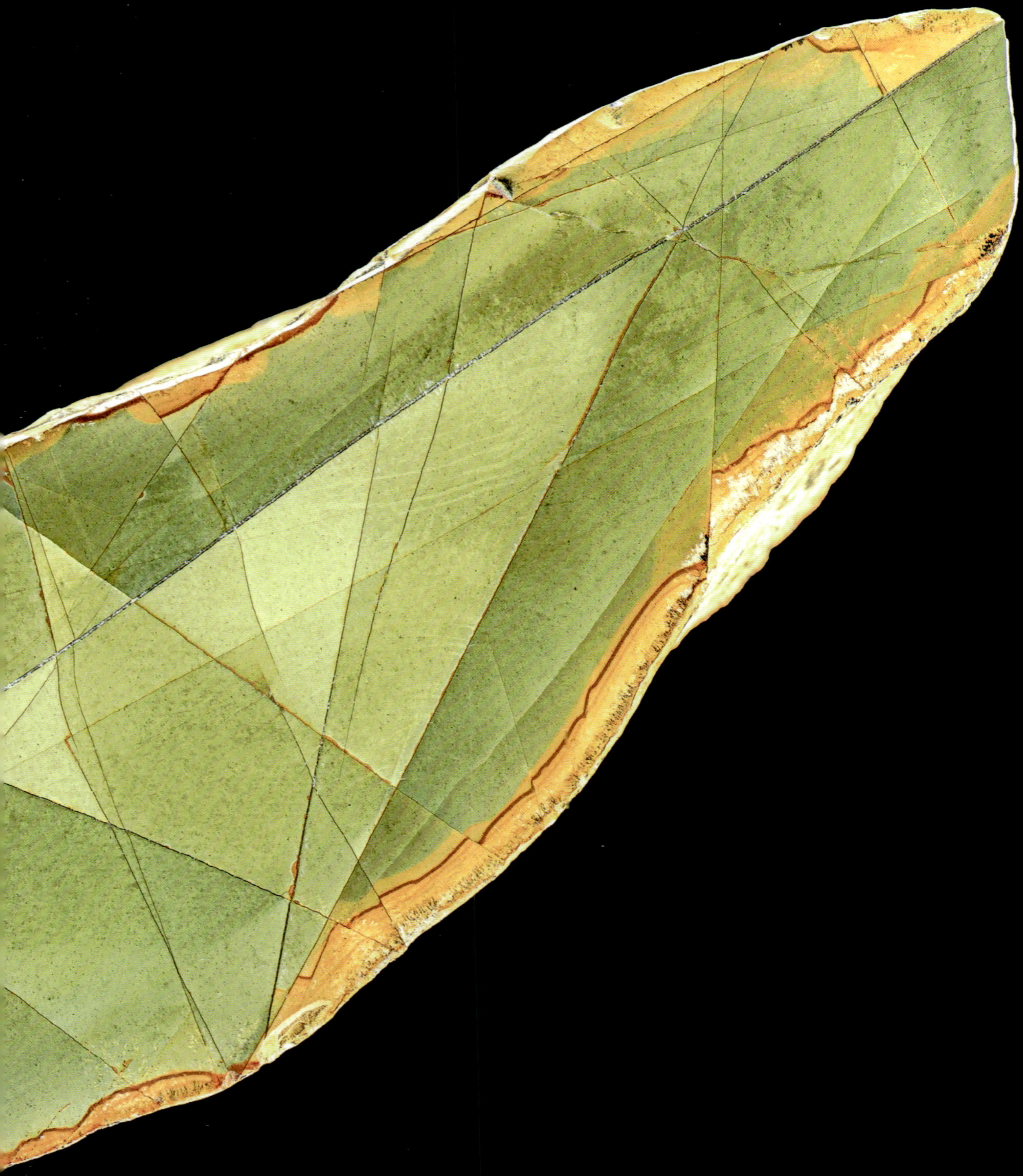

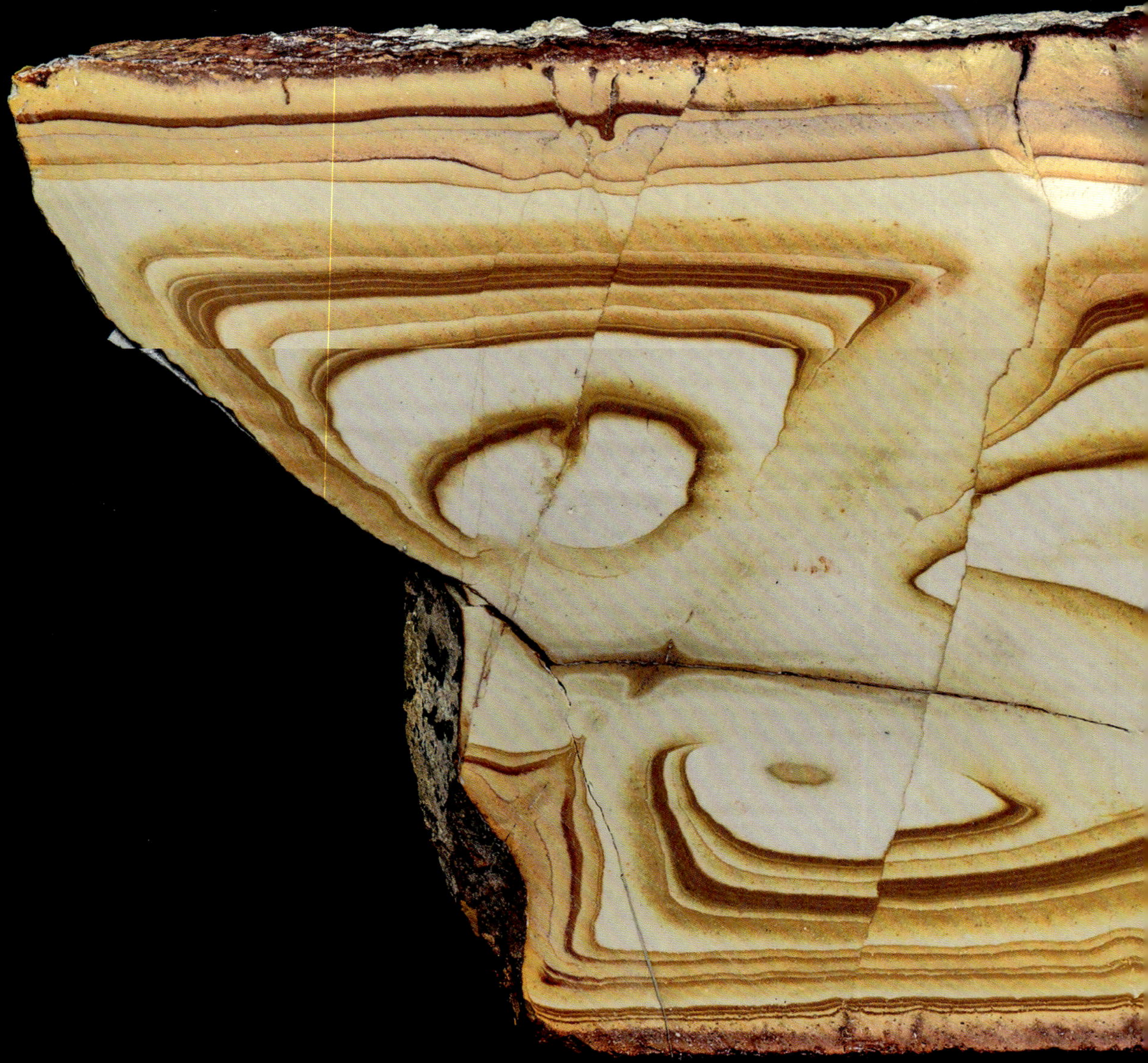

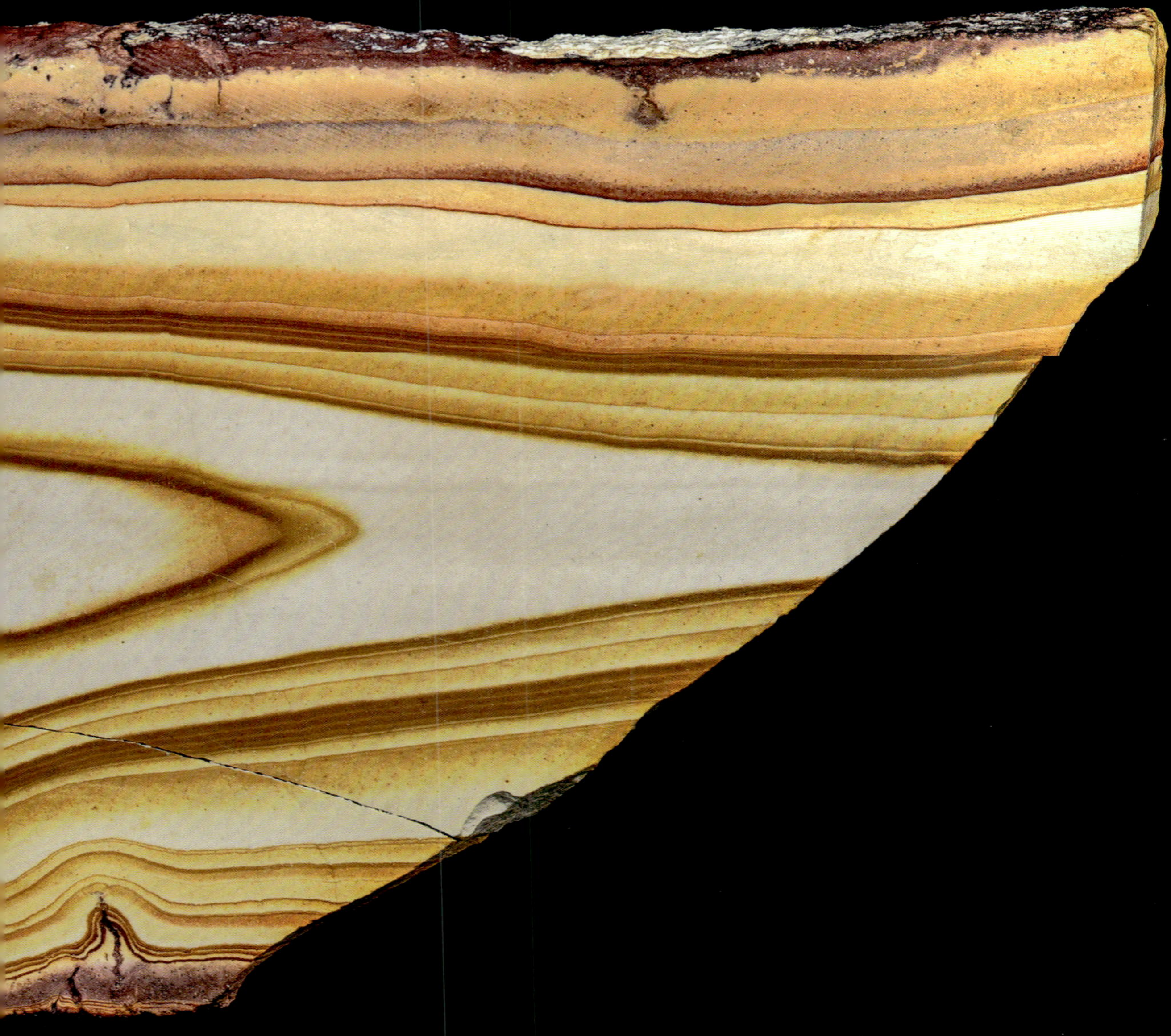

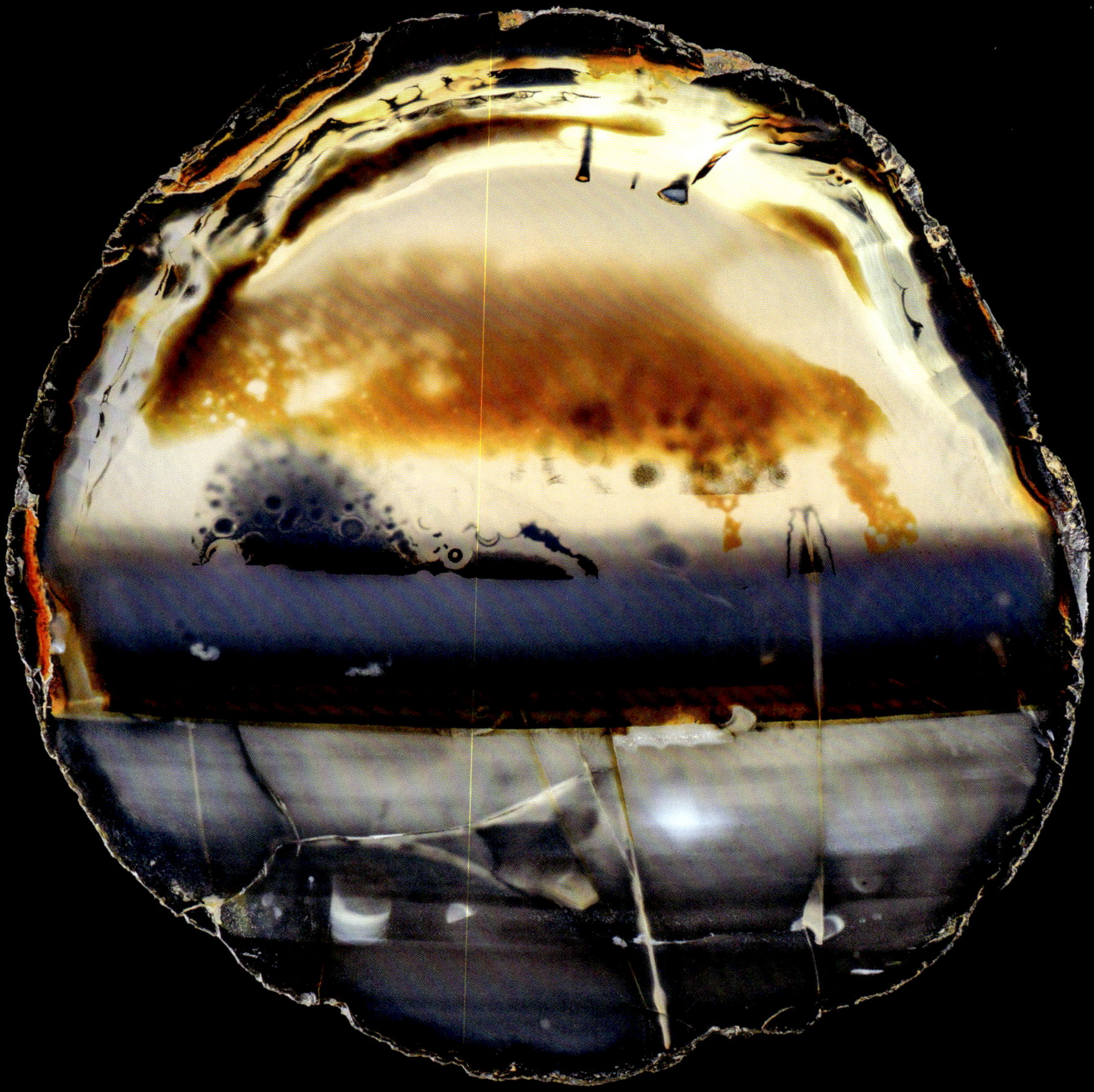

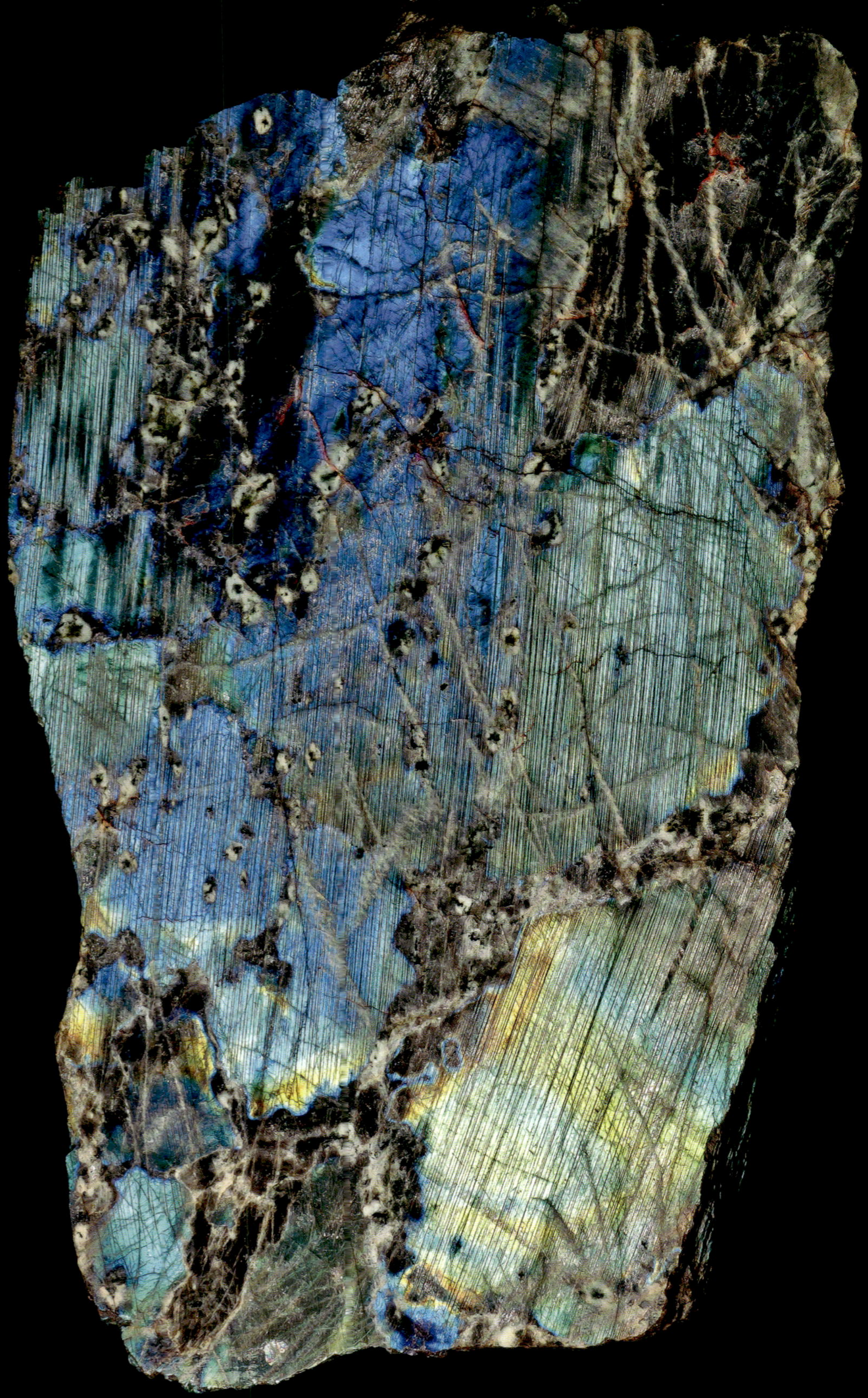

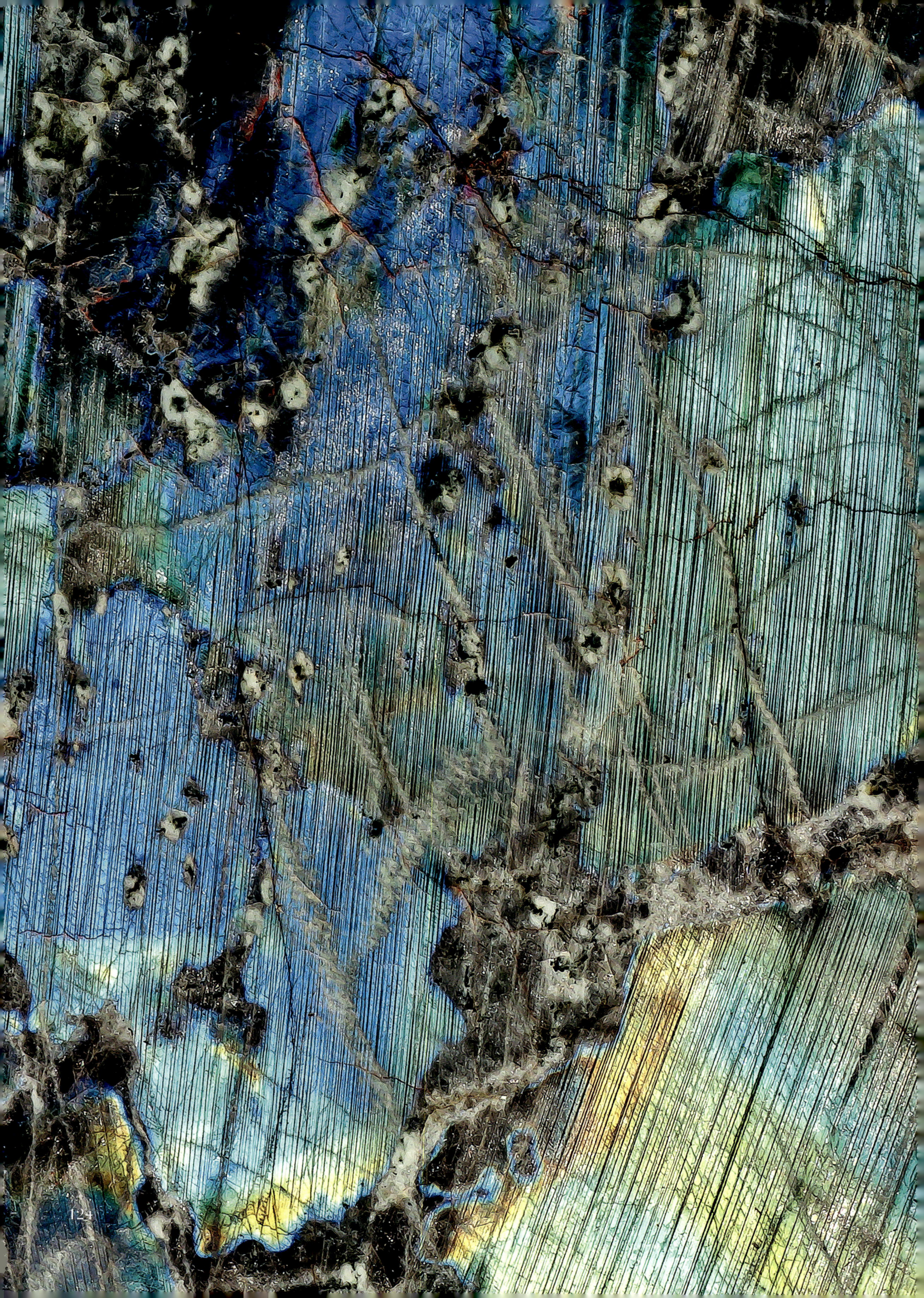

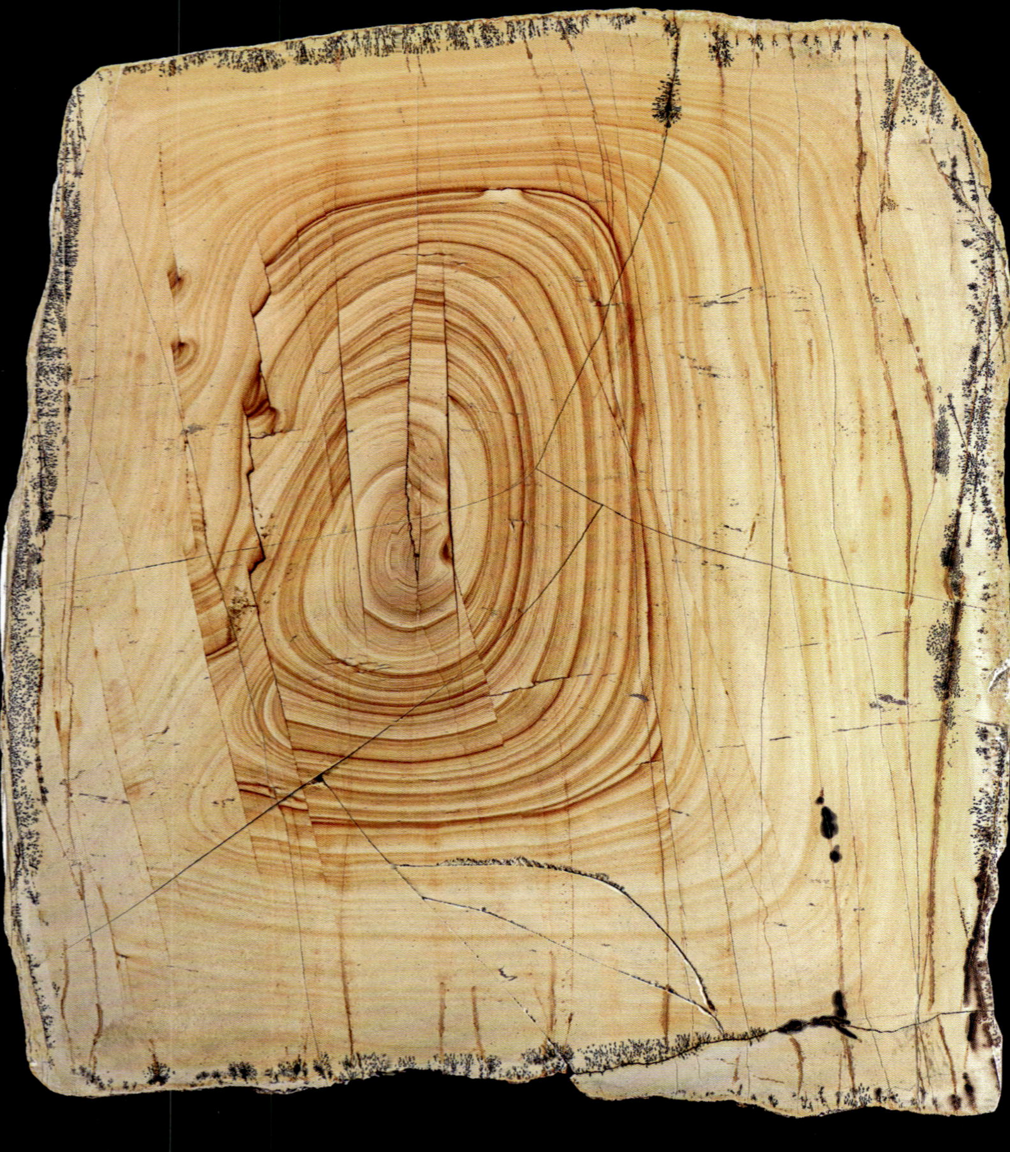

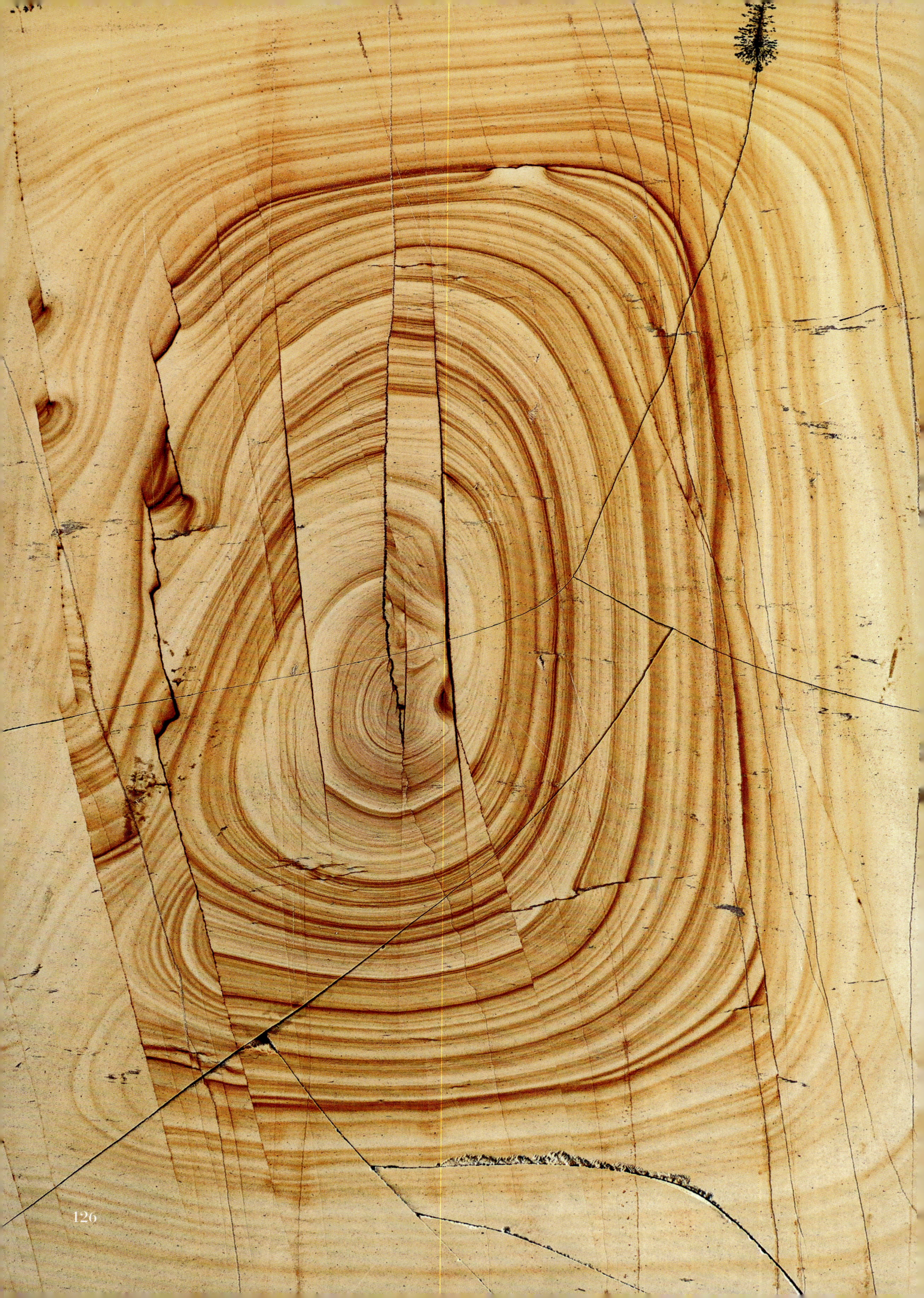

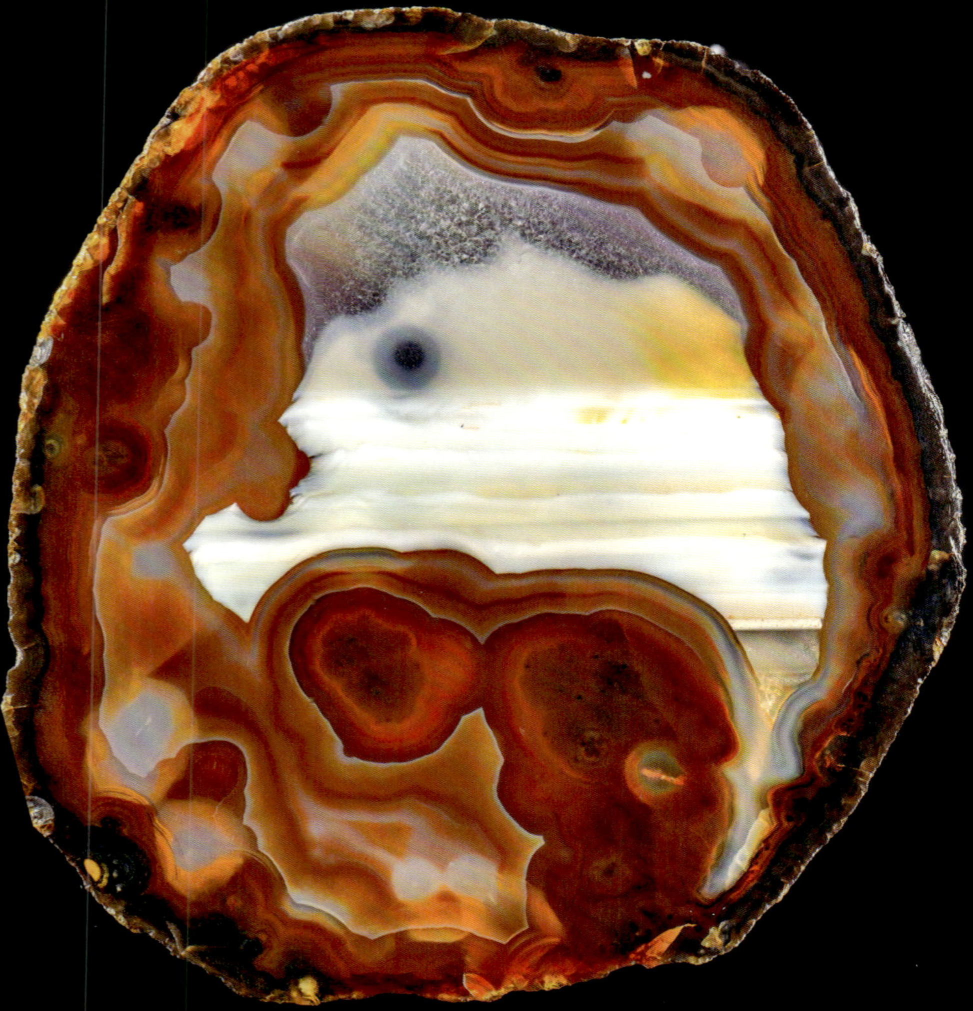

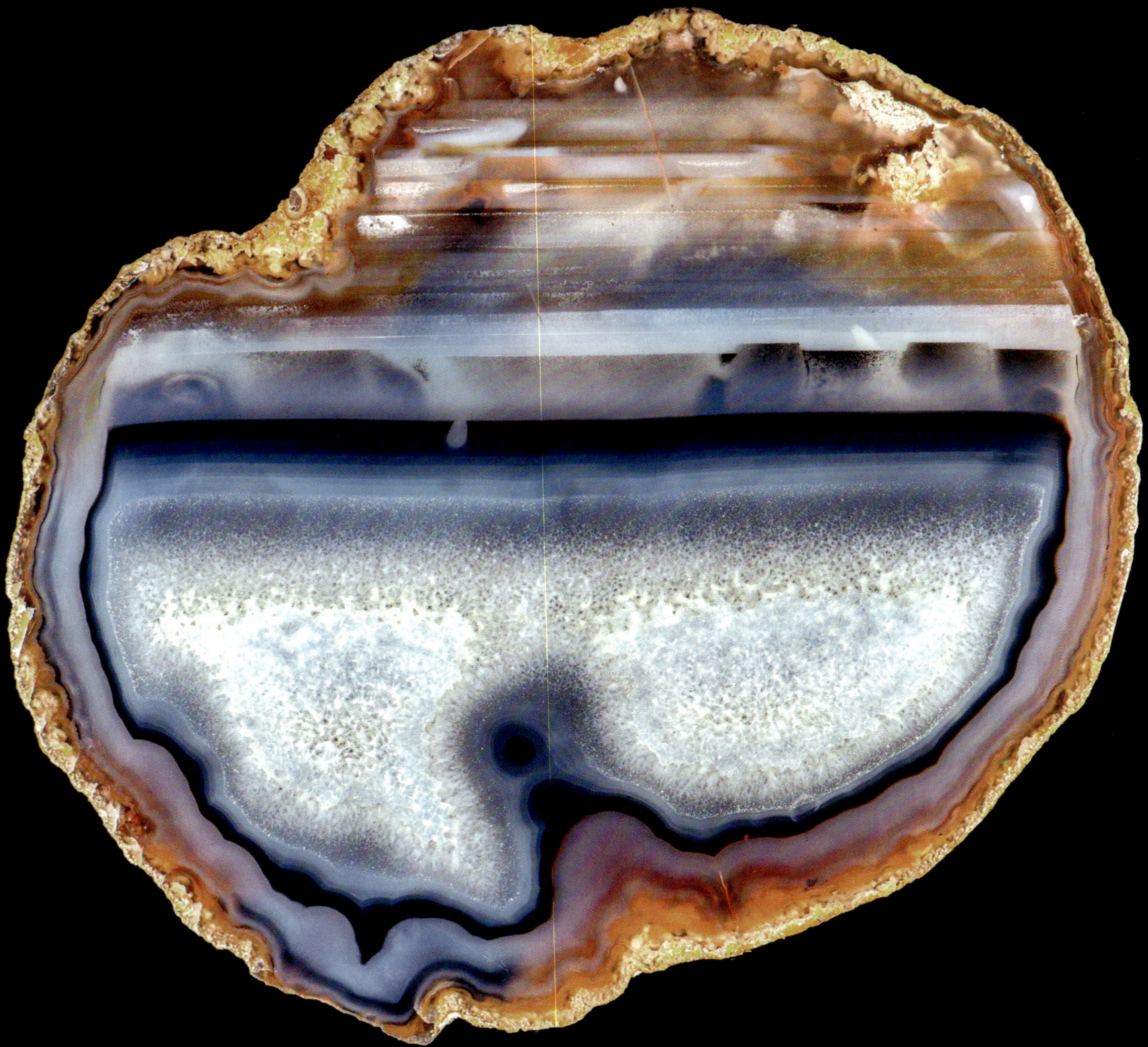

The Man who Loved Stones

by Marguerite Yourcenar

On 22 January 1981, Marguerite Yourcenar officially took the seat of the late Roger Caillois at the Académie Française. She was the first woman in history to be elected to become a member of the Immortals. As was customary, she gave a speech praising her predecessor. What was not customary was the depth, the intimate understanding of the figure of Caillois and the importance in itself of this speech (as well as being a tangible piece of literary prowess), which we have reproduced here in its entirety.

"I gradually gave up regarding man as external to nature, its end;" said Caillois. Far from disparaging what is human, as has been alleged, he found it all along a scale ranging from molecules to the stars. Because he claimed to observe the presence throughout the universe of a sensibility and a consciousness analogous to our own, he has been accused of anthropomorphism. But Caillois himself has argued passionately that on the contrary he advocated an inverted anthropomorphism in which man, instead of attributing his own emotions, sometimes condescendingly, to all other living beings, shares humbly; yet perhaps also with pride, in everything contained or innate in all three realms, animal, vegetable, and mineral. In short, there had taken place in that great intellect the equivalent of the Copernican revolution: man was no longer the center of the universe, except in the sense that the center is everywhere; man, like all the rest, was a cog in the whole system of turning wheels. Quite early on, having entered "the forbidden laboratories," Caillois applied himself to the study of the diagonals which link the species, of the recurrent phenomena that act, so to speak, as a matrix of forms. His work on the octopus and the praying mantis showed him the relation between a creature belonging to the lowest reaches of the animal abyss and the fantasies and desires inhabiting the depths of humanity. In *Méduse et Cie* (Medusa and Company) he meditated on the insect imagination with its extravagant and terrifying transformations, its masks for display or for battle, its nuptial ornaments or instruments of hypnotism, not all of them utilitarian but seemingly corresponding to an almost conscious need for change and elaboration. One of the working hypotheses of modern science – that nature always acts with the greatest possible economy of means and toward the most practical of ends – finally came to appear unacceptable to Caillois. "Nature is not a miser." He had become more aware of it as a mine of prodigality, a feast of superfluity; of the element of fantasy, the aesthetic factor whether unconscious or otherwise which is inherent in every scrap of matter and of which man's own aesthetic may be no more than one of many manifestations, and one often distorted by our exaggerated awareness of it. [...]

Not only in his last works but perhaps even in his earliest writings Caillois shows a kind of indifference toward what is human. This feeling seems to have extended over the whole animal kingdom, with the exception of insects, which are very far removed anatomically and physiologically from our own species, and of creatures like the octopus, traditional objects of fear and horror. Warm-blooded animals, our brothers, seem not to have interested Caillois greatly. Nor does the fish, a distant relation of ours but one we see, when it is snatched from its familiar depths, as a dying creature analogous to a dying human. Apart from the almost fossil dragon trees which, like me, he went to see in the botanic gardens at Orotava, he appears not to have been very much affected even by trees. He liked them best in the form of incorruptible fragments, transformed by millions of centuries during which all that was once juice, sap, and delicate vegetable fiber has been transmuted or cast into amber, agate, or opal, endowed with a mineral, almost eternal, endurance.

But the usual arguments about what is and what is not human prevent us from seeing that in fact Caillois never stops being interested in man. His attitude reminds us, who have so often bored doctors with our clumsy descriptions of symptoms and our (not altogether useless) psychosomatic explanations, of a great specialist consulting his X-rays and chemical analyses and trying to make us understand that the ills attacking us, the death lying in wait for us and the life pulsing through us, exist independently of their physiological signs, and are governed by chemical combinations a thousand miles away from our consciousness and even from our senses. These combinations, separations, and losses, more ancient than our own, Caillois found in the eventful history of stones.

Thus he arrived, though he was rather shy about admitting it, at a "mysticism of matter." I think I can detect in this shyness the effect of two states of mind frequently found in intellectuals of the purely rational type, especially perhaps in France. One is an almost superstitious dread of the word mystic, as if it meant something more than a believer in doctrines that are still more or less secret, or a researcher of matters still hidden. (And yet we all know that all profound thought remains partly secret because of the lack of words to express it, and that everything remains partly hidden from us.) The second of these two states of mind is a certain scorn for the word matter, the latter being too often regarded as substance in. its crudest form, the antithesis of soul not only, as is too often believed, in terms of Christian thought, but also in that of Plato and Aristotle themselves. I would have liked to remind him (though no doubt he did not really forget it) that the pre-Socratic philosophers had preceded him along the same path, and that on the other side of the world Chuang-Tzu would have praised him for having made the transition "from the intelligence which sees differences" – and no one ever did so more subtly than he –"to the intelligence which sees likenesses." David of Dinant, who was burned in Les Halles in the twelfth century, is applauded by Giordano Bruno, who met a similar fate in Rome for "having raised matter to the dignity of something divine." And the Hermetic books tell men to listen to "the great voice of things."

But it is above all when we approach what was to be the supreme object of Caillois's study and love, when we come to the moving books he wrote on stones toward the end of his life, that we hear those distant harmonies. Strangely enough, the symbolism of alchemy compared stone to the human body which, unstable though it may be (and as stone is itself, seen through stretches of time infinitely longer), is yet a "fixed quantity" in comparison with psychic elements, which are even more fluid and mutable. So it is not surprising that instead of choosing gold, which is merely matter transmuted, the alchemists chose the Philosophers' Stone to symbolize transmutation. But let us hear what other great voices have had to say. First, and perhaps above all, the admonition of Jesus in the Apocryphal Gospels: "Split the tree, and I am in the wood; lift a stone, and I am there." More explicit still is Meister Eckhart, one of the greatest mystics of medieval Christendom: "The stone is God, but it does not know it, and it is the not knowing that makes it a stone." And there is also Piranese, who sometimes seems to cherish not so much the antique monument he is ostensibly depicting as the stone from which it was made – stone flaked away by time, devoured by vegetation, forever unaware of the great yet little human events which have taken place around it and left their mark upon it. Goethe was so diligent in the study of stones that one has been named after him (and, hoping that a similar honor may be paid to Caillois, one imagines a list in which the Cailloise figures beside the Goethite); and, when he was growing old, Goethe seems to have enjoyed telling people, "Let the old man play with his stones." The author of *Le Mythe et l'Homme* (Myth and Man) and *L'Homme et le Sacre* (Man and Sacred) makes us think, too, of ancient Mithra, a god born out of the rock. According to what I have been told by a woman who was one of Dag Hammarskjold's friends, that statesman, who was not only an admirer of Saint-John Perse, a poet of whom Caillois also was fond, but also one of the most impressive mystics of our age, set up a sort of oratory in the United Nations building in New York containing only a large block of iron ore in its geological state, with the mineral still there in the original rock. Hammarskjold, a man beset by the ephemeral and recurring conflicts, at once factitious and lethal, of the age of steel and atomic weapons, used to go and conjure up for himself a little silence and serenity in the presence of the ancient block of stone, older than the uses to which it has been put, and still innocent.

Without wishing in the least to compare the two men, of whom one conversed with God right to the end while the other concentrated on the immanence hidden in the depths of things, no reader of *Pierres réfléchies* (Stone Reflections), *Recurrences dérobées* (Hidden Recurrences) and above all *Le Fleuve Alphée* (The River Alpheus) can doubt that Roger Caillois, like so many of us, felt terribly weary when he contemplated the restlessness of modern mankind and the world wide upheavals it has produced. Man's is an abnormal and "therefore precarious" situation. The future is dark. "Through knowledge and genius, man has succeeded in drawing on the energy in the nucleus of the fundamental particles that contain nature's fundamental reserves: and it is not improbable that a chain reaction, imperfectly controlled or carelessly provoked, may liberate an excessive quantity of that energy and annihilate all matter. The intersecting paths of Chance and Necessity have presided over man's prodigious destiny; but they also suggest that the miracle might happen in reverse, and return life to the impassive and immortal inertia from which a lucky statistical chance once plucked it." Caillois's devotion and emotion seem to recoil from a humanity conceived of as more precarious than ever, and from an animal and vegetable kingdom whose destruction we are hastening; he seeks a substance more lasting, an object more pure. And he finds it in the race of stones: "the dim mirror of obsidian;" vitrified thousands of centuries ago at temperatures no longer known; the diamond, which while still buried in the earth already carries within it all the potentialities of its future fires; the evanescence of mercury; crystal, giving lessons to man in advance by admitting impurities which endanger its transparency and balance – splinters of iron, tufts of chlorite, threads of rutile – yet, in spite of all these, pursuing its limpid growth, with prisms which, as Caillois brilliantly reminds us, are like souls in casting no shadows. Not only did the amazing variety of form found in stones persuade Caillois that human invention is only a development of the data inherent in things, but also in minerals through aesthetics, he found history. Those fusions, pressures, ruptures, imprints of matter on matter have left traces inside and out which sometimes almost exactly resemble writing and which actually do transcribe events from millions of years ago. "There are impossible scribblings in nature, written neither by men nor by devils:" and seeming to foreshadow the insatiable human passion for meaning and recording. "Already present in the archives of geology, available for operations then inconceivable, was the model of what would later be an alphabet." Although, as Caillois knew better than anyone, this unconscious alphabet is immeasurably far away from the lines of letters we produce by a turn of the wrist, itself the slave of muscles, tendons, and neurons, yet the authorless inscriptions may be regarded as a first draft of a chronicle of stones. Stones, like us, stand at the intersection of countless lines crossing one another and receding to infinity, at the center of a field of forces too unpredictable to be measured; and we awkwardly call the result chance, hazard, or fate.

Index of plates

Page	*Name*	*Origin*	*Cut*
6	Rhodocrosite	Minas Capilitas (Andalgalá, Catamarca, Argentina)	Thick segment polished in a stalactite
17, 58	Agate	Probably Rio Grande do Sul (Brazil)	Thin slab sawn and polished
18/19	*Broussailles*, limestone with birnessite dendrites, Tuscan variety known as lineato	Upper Arno Valley (Florence, Italy)	Thin cross-section sawn and polished
20/21	Agate	Probably Rio Grande do Sul (Brazil)	Slab, sawn and polished
22	Sycamore fossile wood (Miocene, 15/16 million years ago)	Rough Badger Pocket (Washington, USA)	Seminodular, with polished surface
23	*Le fantôme*, agate	Probably Rio Grande do Sul (Brazil)	Thin cross-section sawn and polished
24	*Calligraphie royale*, onyx	Probably Rio Grande do Sul (Brazil)	Block with polished surface
25-27	*Quartz variety rock crystal with onyx "eyes."*	Artigas (Uruguay)	Slab, sawn and polished
28-31	Eye agate, fragment	Uruguay	Thick segment, sawn and polished
32	Eye agate in quartz, rock crystal	Artigas (Uruguay)	Thick cross-section polished on one side
41	*Oakstone*, baryte	Arbor Low (Derbyshire, UK)	Thick concretion polished on one side
49	*Agate blessée*	Probably Rio Grande do Sul (Brazil)	Thick segment, sawn and polished
50	Polyhedric agate with birnessite dendrites	Sítio Garguelo (Paraíba, Brazil)	Half geode sawn and polished
51, 52	Graphic limestone	Upper Arno Valley (Florence, Italy)	Block polished on one side
53	Slate shale with fossile of pyritized ophiura (Devonian)	Budenbach (Rhineland-Palatinate, Germany)	
54	Variscite	Little Green Monster mine (Fairfield, Utah, USA)	Seminodular polished with carnallite and cardite
55	Liddicoatite (tourmaline family)	Madagascar	Thick cross-section, polished
56/57	Amethyst in chalcedony	Probably Artigas (Uruguay)	Thick segment, sawn and polished
59	Polyhedric agate	Sítio Garguelo (Paraíba, Brazil)	Thin cross-section, sawn and polished
60	Chalcedony	Rio Grande do Sul (Brazil)	Seminodular, sawn and polished
61	*L'œil bleu*, agate and quartz	Probably Rio Grande do Sul (Brazil)	Thick cross-section, polished
62/63	*La rature*, agate	Probably Rio Grande do Sul (Brazil)	Thin cross-section, sawn and polished
64	*Le monstre*, quartz and chalcedony	USA	Nodule, sawn in half and polished
77, 78	*La cime*, paesine agate	Chichuahua (Mexico)	Nodule, sawn and polished
79	Rhodonite with black birnessite dendrites	Ampanihy (Atsimo-Andrefana, Madagascar)	Thick segment, sawn and polished
80/81	Pierre aux-masures, clay limestone, Tuscan variety known as *paesina*	Upper Arno Valley (Florence, Italy)	Thin cross-section, sawn and polished

All of the stones in the Caillois collection featured in this volume are now at the Muséum national d'histoire naturelle in Paris.

Dimensions	*Provenance*	*Inventory*	*Notes*
15 x 19 cm	Caillois Donation, 1984-1985	MIN008-039	
16,6 x 9,9 cm	Gift École Van Cleef & Arpels des Arts Joailliers, 2017	Gg-2021-6236	
24,5 x 8 cm	Gift École Van Cleef & Arpels des Arts Joailliers, 2017	Gg-2021-5107	
38 x 17 cm	Caillois Donation, 1984-1985	188.98	Inventoried as *Le Vaisseau* probably an apocryphal name
14,5 x 9 cm	Gift École Van Cleef & Arpels des Arts Joailliers, 2017	Gg-2021-6501	
19 x 23 cm	Caillois Donation, 1984-1985	188.117	
14 x 10 cm	Gift École Van Cleef & Arpels des Arts Joailliers, 2017	Gg-2021-6243	
23 x 25 cm	Caillois Donation, 1984-1985	188.105	
15 x 13 cm	Caillois Donation, 1984-1985	MIN008-152	Inventoried as *Monocle et binocles* and later *Oeil et monocle*, probably apocryphal names
18 x 25 cm	Caillois Donation, 1984-1985	188.118	
23 x 29 cm	Caillois Donation, 1984-1985	188.126	
30 x 34 cm	Caillois Donation, 1984-1985	188.77	
11,5 x 5,7 cm	Gift École Van Cleef & Arpels des Arts Joailliers, 2017	Gg-2021-5114	
25,7 x 29,3 cm	Gift École Van Cleef & Arpels des Arts Joailliers, 2017	Gg-2021-5099	
9,2 x15,4 cm	Gift École Van Cleef & Arpels des Arts Joailliers, 2017	Gg-2021-7528	
17 x 19 cm	Donazione Callois, 1984-1985	188.62	
24 x 27 cm	Caillois Donation, 1984-1985	188.58	Inventoried as *La Masque*
19,2 x 6,7 cm	Gift École Van Cleef & Arpels des Arts Joailliers, 2017	Gg-2021-6224	
11,7 x 14,4 cm	Caillois Donation, 1984-1985	188.166	
9,2 x 9,5 cm	Gift École Van Cleef & Arpels des Arts Joailliers, 2017	Gg-2021-6252	
15 x 10 cm	Caillois Donation, 1984-1985	188.101	
19 x 13 cm	Caillois Donation, 1984-1985	188.140	
12,6 x11,2 cm	Gift École Van Cleef & Arpels des Arts Joailliers, 2017	Gg-2021-6307	
9 x 9,5 cm	Gift École Van Cleef & Arpels des Arts Joailliers, 2017	Gg-2021-5882	
28 x 27 cm	Caillois Donation, 1984-1985	INV. 188.124	
22,8 x 9,8 cm	Gift École Van Cleef & Arpels des Arts Joailliers, 2017	Gg-2021-5070	

82/83	*Paysage*, limestone with stromatoliths from the Rhaetian Age	Cotham (Bristol, England)	Thick slab, sawn and polished
84	*Entrée de grotte*, Pierre aux-masures, clay limestone, Tuscan variety known as *paesina*	Upper Arno Valley (Florence, Italy)	Half block sawn and polished
85	*Monstre décharné*, septarian	Black Rock Desert (Nevada, USA)	Seminodular, sawn and polished
86/87	Septarian	Probably Deba-Zumaia (Gipuzkoa, Euskadi, Spain)	Seminodular, sawn and polished
88/89	Septarian	Probably Deba-Zumaia (Gipuzkoa, Euskadi, Spain)	Seminodular, sawn and polished
90	Septarian	Otzenhausen (Sarre, Germany)	Seminodular, sawn and polished
91	Septarian		Seminodular, sawn and polished
92	Agate with curvilinear diamond shape	Rio Grande do Sul (Brazil)	Thin cross-section, sawn and polished
93	*Calligraphie*, onyx (front and back)	Probably Rio Grande do Sul (Brazil)	Thin cross-section, sawn and polished
94	*Arc-au-ciel, iridescent obsidian*	Probably La Revoltosa (Jalisco, Mexico)	Thick slab, sawn and polished on one side
95	*Oiseau Naissant*, agate, on flint	Probably Rio Grande do Sul (Brazil)	Very thick cross-section
96	*Orbites*, jasper	Bruneau (Idaho, USA)	Thick segment, sawn and polished
97	*La fissure*, onyx	Probably Rio Grande do Sul (Brazil)	Block, sawn and polished on one side
98	Jasper	Deschutes County (Oregon, USA)	Thin slab, sawn and polished
99, 100	Scaled jasper	Madagascar	Thin cross-section, sawn and polished
113	*Le château*, limestone with dendrites	Probabilmente Baden-Württemberg (Germania)	Lithographic limestone sawn, polished then micro-engraved
114	*Le portrait*, limestone with birnessite dendrites, Tuscan variety known as *lineato*	Upper Arno Valley (Florence, Italy)	Thin cross-section, sawn and polished
115	Entrée de la vie, *limestone with belemnite rostra, Lower Jurassic*	Paris (France)	Thick segment, sawn and polished
116	Limestone, Tuscan variety known as *Verde d'Arno*	Upper Arno Valley (Florence, Italy)	Very thin cross-section, sawn and polished
117	Sepiolite carved, polished and mounted on ebony	Killik (Eskisehir, Turchia)	
118/119	Limestone, Tuscan variety known as *Verde d'Arno*	Upper Arno Valley (Florence, Italy)	Block sawn and polished
120/121	*Visage anamorphique*, jasper	USA	Thick segment, sawn and polished
122	Agate	Probably Rio Grande do Sul (Brazil)	Thin cross-section, sawn and polished
123/124	Labradorite, variety of iridescent anorthite	Ylämaa (South Karelia, Finland)	Thick slab, sawn and polished on one side
125/126	*Diaspre de l'Arno*, limestone with birnessite dendrites, Tuscan variety known as *lineato*	Upper Arno Valley (Florence, Italy)	
127	*Larve*, chalcedony	Probably Rio Grande do Sul (Brazil)	Seminodular, sawn and polished
128	Agate	Probably Rio Grande do Sul (Brazil)	Thin cross-section, sawn and polished

28 x 8 cm	Gift École Van Cleef & Arpels des Arts Joailliers, 2017	Gg-2021-5125	Caillois wronlgy inventoried it as from Catham
8 x 12 cm	Gift École Van Cleef & Arpels des Arts Joailliers, 2017	Gg-2021-5057	
12 x 11 cm	Gift École Van Cleef & Arpels des Arts Joailliers, 2017	188.64	
12,3 x 7,1 cm	Gift École Van Cleef & Arpels des Arts Joailliers, 2017	Gg-2021-5302	
11,1 x 7,3 cm	Gift École Van Cleef & Arpels des Arts Joailliers, 2017	Gg-2021-5301	
9,6 x 9 cm	Gift École Van Cleef & Arpels des Arts Joailliers, 2017	Gg-2021-5735	
7,5 x 7,0 cm	Gift École Van Cleef & Arpels des Arts Joailliers, 2017	Gg-2021-5741	
11 x 16 cm	Caillois Donation, 1984-1985	188.66	
14 x 12 cm	Caillois Donation, 1984-1985	188.63	
25 x 22 cm	Caillois Donation, 1984-1985	188.125	
15 x 23 cm	Caillois Donation, 1984-1985	188.100	Applied by Caillois on a marine stone
3 x 9 cm	Donazione Callois, 1984-1985	188.93	
10 x 15 cm	Caillois Donation, 1984-1985	188.102	
26 x 21 cm	Caillois Donation, 1984-1985	188.113	
16,3 x 19,3 cm	Gift École Van Cleef & Arpels des Arts Joailliers, 2017	Gg-2021-5908	
6 x 5 cm	Gift École Van Cleef & Arpels des Arts Joailliers, 2017	Gg-2021-3065	
9 x 13,5 cm	Gift École Van Cleef & Arpels des Arts Joailliers, 2017	Gg-2021-5092	
20 x 13 cm	Caillois Donation, 1984-1985	188.59	
14 x 11 cm	Gift École Van Cleef & Arpels des Arts Joailliers, 2017	Gg-2021-5091	
10 x 23 cm	Caillois Donation, 1984-1985	199.323	
15,1 x 5,9 cm	Caillois Donation, 1984-1985	188.94	
17 x 7 cm	Gift École Van Cleef & Arpels des Arts Joailliers, 2017	Gg-2021-5051	
15 x 16 cm	Caillois Donation, 1984-1985	188.127	
23 x 29,5 cm	Caillois Donation, 1984-1985	N.C.	
19 x 21 cm	Caillois Donation, 1984-1985	188.67	
9,5 x 10 cm	Gift École Van Cleef & Arpels des Arts Joailliers, 2017	Gg-2021-5807	
27,2 x 24,7 cm	Gift École Van Cleef & Arpels des Arts Joailliers, 2017	Gg-2021-6240	

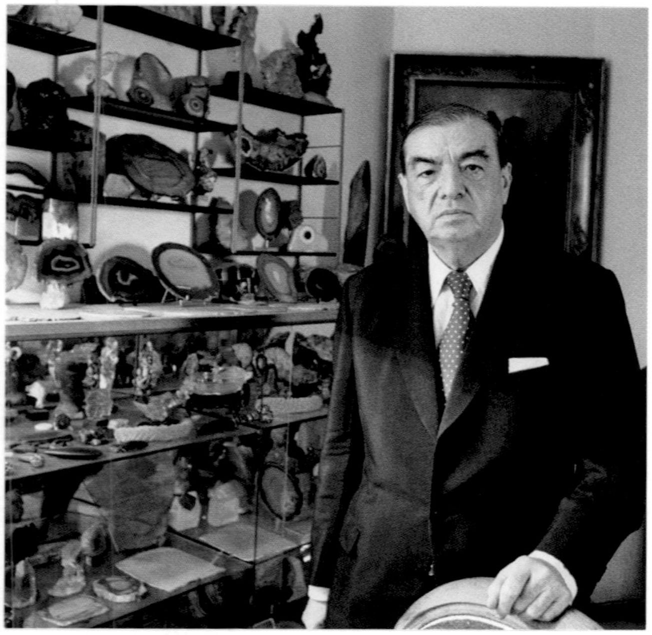

This volume
was printed in Milan
by Grafiche Milani
for Franco Maria Ricci editore
in November 2023